电气工程、自动化专业规划教材

电气工程及其自动化专业英语
（第2版）

戴文进　编著

电子工业出版社

Publishing House of Electronics Industry

北京·BEIJING

内 容 简 介

本书从适应高等学校专业英语的教学需要出发，较全面涵盖了电气工程及其自动化专业英语的相关知识。全书共 5 大部分，分别为：电工电子、电机及其控制、发电输电配电与电力系统、可再生能源发电、计算机与智能技术，每一部分由英语原文、专业英语词汇、课文注释、参考译文组成。

本书取材新颖、内容丰富、注释详尽，是电气工程及其自动化专业英语的适用教材，也可作为其他相近专业的参考书，还可供有关技术人员选用。

未经许可，不得以任何方式复制或抄袭本书之部分或全部内容。
版权所有，侵权必究。

图书在版编目(CIP)数据

电气工程及其自动化专业英语 / 戴文进编著. —2 版. —北京：电子工业出版社，2018.5
电气工程、自动化专业规划教材
ISBN 978-7-121-34127-4

Ⅰ. ①电… Ⅱ. ①戴… Ⅲ. ①电气工程－英语－高等学校－教材②自动化技术－英语－高等学校－教材
Ⅳ. ①TM②TP1

中国版本图书馆 CIP 数据核字(2018)第 087105 号

责任编辑：凌　毅
印　　刷：涿州市京南印刷厂
装　　订：涿州市京南印刷厂
出版发行：电子工业出版社
　　　　　北京市海淀区万寿路 173 信箱　邮编　100036
开　　本：787×1 092　1/16　印张：13.25　字数：450 千字
版　　次：2011 年 3 月第 1 版
　　　　　2018 年 5 月第 2 版
印　　次：2024 年 6 月第 13 次印刷
定　　价：39.00 元

凡所购买电子工业出版社图书有缺损问题，请向购买书店调换。若书店售缺，请与本社发行部联系，联系及邮购电话：(010) 88254888，(010) 88258888。
质量投诉请发邮件至 zlts@phei.com.cn，盗版侵权举报请发邮件至 dbqq@phei.com.cn。
本书咨询联系方式：(010) 88254528，lingyi@phei.com.cn。

再 版 前 言

近年来，我国电气工程及其自动化行业迎来了新一轮的发展机遇，但也面临着国外电气工程及其自动化先进技术和高品质产品的严峻挑战。随着世界经济一体化进程的加快，我国与国外同行之间的业务交往日益增多，大量的先进电气工程设备的引进也随之增多，电气工程及其自动化专业英语的应用越来越广泛。对于电气工程及其自动化专业的大学生和从事相关领域的工程技术人员来讲，熟练掌握电气工程及其自动化专业英语这一基本技能是非常必要的。

本书第 1 版自 2011 年出版以来，深受广大读者的欢迎和大力支持，该书的重印量不断攀升。但是，现代电气工程及其自动化技术发展迅速，该书毕竟出版时间较长，其中不少内容已显陈旧，因此有必要进行重新修订。

本书选材于国内外出版的教科书、专著、外文期刊等。尽管所选内容篇幅有限，但比较精练，基本涵盖了电气工程及其自动化学科领域的专业基础知识。全书共 5 大部分，分别为：电工电子、电机及其控制、发电输电配电与电力系统、可再生能源发电、计算机与智能技术，每一部分由英语原文、专业英语词汇、课文注释、参考译文组成。本书取材新颖、内容丰富、注释详尽，是电气工程及其自动化专业英语的适用教材，也可作为其他相近专业的参考书，还可供有关技术人员选用。

本书由戴文进教授主持编写，虽然长期工作在专业英语的教学第一线，且对该门课程的教学内容和方法有一定体会，但毕竟水平有限，故书中谬误之处在所难免，敬请读者不吝指正。

戴文进
2018 年 4 月

目　录

Part 1　Electrics & Electronics ··· 1

Unit 1.1　Elements and Parameters in a Circuit ·· 2
Unit 1.2　Ideal Sources Series and Parallel Equivalent Circuits ·························· 8
Unit 1.3　Analysis of Sinusoidal Alternating Electricity ································· 14
Unit 1.4　Analysis of Small Signal Amplifiers in Mid-Frequency Band ············· 22
Unit 1.5　Logic Circuits ·· 29

Part 2　Electric Machine & Its Control ·· 39

Unit 2.1　Transformer Principle ··· 40
Unit 2.2　D.C. Machine Construction and Principle ······································· 48
Unit 2.3　Three-Phase Induction Machine Principle ······································· 58
Unit 2.4　Operation Analysis of Three-Phase Synchronous Generator ·············· 67
Unit 2.5　Motor Control Based on Microcomputer ·· 75

Part 3　Electric Generation, Distribution & Power System ····························· 83

Unit 3.1　Major Electrical Plant in Power Station ·· 84
Unit 3.2　Operation & Control of Power System ··· 97
Unit 3.3　Relaying Protection in Power System ··· 104
Unit 3.4　SCADA & EMS in Power System ·· 112
Unit 3.5　Electric Power Flow Calculation in Power System ························ 120

Part 4　Electric Generation with Renewable Energy ···································· 129

Unit 4.1　Global Status of Wind Energy Market ·· 130
Unit 4.2　Development on Electric Generation With Nuclear Power ·············· 135
Unit 4.3　Ecological Protection Project of Small Hydroelectric Power Replace of Fuel ······ 139
Unit 4.4　Discussion on Biomass Power Generation ···································· 146
Unit 4.5　Hybrid PV-Battery-Diesel Power System ······································ 155

Part 5　Computer & Artificial Intelligence ··· 161

Unit 5.1　Computer and Its Hardware ··· 162
Unit 5.2　Computer Software and Human Resources ··································· 169

Unit 5.3　Personal Computer Word Processing and Electronic Spreadsheets ········· 180

Unit 5.4　Intelligent Technology ·· 191

Unit 5.5　Neural Nets ·· 198

Reference ··· 205

Part 1   Electrics & Electronics

Unit 1.1   Elements and Parameters in a Circuit

Unit 1.2   Ideal Sources Series and Parallel Equivalent Circuits

Unit 1.3   Analysis of Sinusoidal Alternating Electricity

Unit 1.4   Analysis of Small Signal Amplifiers in Mid-Frequency Band

Unit 1.5   Logic Circuits

Unit 1.1 Elements and Parameters in a Circuit

1.1.1 Text

An electric circuit (or network) is an interconnection of physical electrical devices. The purpose of electric circuits is to distribute and convert energy into some other forms. Accordingly, the basic circuit components are an energy source (or sources), an energy converter (or converters), and conductors connecting them.

An energy source (a primary or secondary cell, a generator, and the like) converts chemical, mechanical, thermal or some other form of energy into electric energy. An energy converter, also called load (such as a lamp, heating appliance, or electric motor), converts electric energy into light, heat, mechanical work, and so on.

Events in a circuit can be defined in terms of e.m.f. (or voltage) and current. When electric energy is generated, transmitted and converted under conditions such that the currents and voltages involved remain constant with time, one usually speaks of direct-current (D.C.) circuits【1】.

With time-invariant currents and voltages, the magnetic and electric fields of the associated electric plant are also time-invariant. This is the reason why no e.m.f.s of self-or mutual-induction appear in D.C. circuits, nor are there any displacement currents in the dielectric surrounding the conductors.

Fig.1 (omitted) shows in simplified form a hypothetical circuit with a storage battery as the source and a lamp as the load. The terminals of the source and load are interconnected by conductors (generally but not always wires)【2】. As is seen, the source, load and conductors form a closed conducting path. The e.m.f. of the source causes a continuous and unidirectional current to circulate round this closed path.

This simple circuit made up of a source, a load and two wires is seldom, if ever, met with in practice. Practical circuits may contain a large number of sources and loads interconnected in a variety of ways【3】.

To simplify analysis of actual circuits, it is usual to show them symbolically in a diagram called a circuit diagram, which is in fact a fictitious or, rather, idealized model of an actual circuit of network. Such a diagram consists of interconnected symbols called circuit elements or circuit parameters. Two elements are necessary to represent processes in a D.C. circuit. These are a source of e.m.f. E and of internal (or source) resistance R_s, and the load resistance (which includes the resistance of the conductors) R (Fig.2, omitted).

Whatever its origin (thermal, contact, etc.), the source e.m.f. E (Fig.2 (a), omitted) is numerically equal to the potential difference between terminals 1 and 2 with the external circuit open【4】, that is, when there is no current flowing through the source:

$$E=\varphi_1-\varphi_2=V_{12} \tag{1}$$

The source e.m.f. is directed from the terminal at a lower potential to that at a higher one. On diagram, this is shown by arrows.

When a load is connected to the source terminals (the circuit is then said to be loaded) and the circuit is closed, a current begins to flow round it. Now the voltage between source terminals 1 and 2 (called the terminal voltage) is not equal to its e.m.f. because of the voltage drop V_s inside the source, that is, across the source resistance R_s,

$$V_s = R_s I$$

Fig.3 (omitted) shows a typical so-called external characteristic $V = \varphi_1 - \varphi_2 = V(I)$ of a loaded source (hence another name is the load characteristic of a source). As is seen, increase of current from zero to $I \approx I_1$ causes the terminal voltage of the source to decrease linearly

$$V_{12}=V=E-V_s=E-R_s I$$

In other words, the voltage drop V_s across the source resistance rises in proportion to the current. This goes on until a certain limit is reached. Then as the current keeps rising, the proportionality between its value and the voltage drop across the source is upset, and the external characteristic ceases to be linear. This decrease in voltage may be caused by a reduction in the source voltage, by an increase in the internal resistance, or both.

The power delivered by a source is given by the equality

$$P_s=EI \tag{2}$$

where P_s is the power of the source.

It seems relevant at this point to dispel a common misconception about power. Thus one may hear that power is generated, delivered, consumed, transmitted, lost, etc. in point of fact, however, it is energy that can be generated, delivered, consumed, transmitted or lost. Power is just the rate of energy input or conversion, that is, the quantity of energy generated, delivered, transmitted, etc. per unit time. So, it would be more correct to use the term energy instead of power in the above context. Yet, we would rather fall in with the tradition.

The load resistance R(Fig.2(b), omitted), as a generalized circuit element, gives an idea about the consumption of energy, that is, the conversion of electric energy into heat, and is defined as

$$P=RI^2 \tag{3}$$

In the general case, the load resistance depends solely on the current through the load, which fact is symbolized by the function $R(I)$.

By Ohm's law, the voltage across a resistance is

$$V=RI \tag{4}$$

In circuit analysis, use is often made of the reciprocal of the resistance, termed the conductance, which is defined as

$$g=1/R$$

In practical problems, one often specifies the voltage across a resistance as a function of current, $V(I)$, or the inverse relation $I(V)$ have come to be known as volt-ampere characteristics.

Fig.4 (omitted) shows volt-ampere curves for a metal-filament lamp, $V_1(I)$, and for a carbon-filament lamp $V_2(I)$. As is seen, the relation between the voltage and the current in each lamp is other than linear. The resistance of the metal-filament lamp increases and that of the carbon-filament lamp decreases with increase of current.

Electric circuits containing components with non-linear characteristic are called non-linear.

If the e.m.f and internal resistances of sources and associated load resistances are assumed to be independent of the current and voltage, respectively, the external characteristic $V(I)$ of the sources and the volt-ampere characteristic $V_1(I)$ of the loads will be linear（Fig.5，omitted）.

Electric circuits containing only elements with linear characteristic are called linear.

Most practical circuits may be classed as linear. Therefore, a study into the properties and analysis of linear circuits is of both theoretical and applied interest.

1.1.2 Specialized English Words

circuit components   电路元件
circuit parameters   电路参数
electric circuit   电路
electrical device   电气设备
electric energy   电能
energy source   电源
primary cell   原生电池
secondary cell   再生电池
energy converter   电能转换器
conductor   导体
generator   发电机
heating appliance   电热器
direct-current（D.C.）circuit   直流电路
magnetic and electric field   电磁场
time-invariant   时不变的
self-（or mutual-）induction   自（互）感
displacement current   位移电流
the dielectric   电介质
storage battery   蓄电池
wire   导线
e.m.f.＝electromotive force   电动势
unidirectional current   单方向电流
circuit diagram   电路图
load characteristic   负载特性
terminal voltage   端电压
external characteristic   外特性
load resistance   负载电阻
voltage drop   电压降

conductance　电导
volt-ampere characteristics　伏安特性
metal-filament lamp　金属丝灯泡
carbon-filament lamp　碳丝灯泡
non-linear characteristics　非线性特性

1.1.3　Notes

【1】When electric energy is generated, transmitted and converted under conditions such that the currents and voltages involved remain constant with time, one usually speaks of direct-current (D.C.) circuits.

这是一个主从复合句，when...time，是一个由 when 引导的时间状语从句，one...circuits. 是整个句子的主句。只是在从句中又包含了一个由 such that 引导的结果状语从句，该从句的主语是 the currents and voltages，involved 是它的后置定语，意为"牵涉的，涉及的"，remain 是系动词做谓语，结果状语从句是系表结构。因此，整句翻译成"当电能在产生、传输和变换时，若电路中相关的电流和电压不随时间而变化，我们便称其为直流电路。"可以看到，结果状语从句并未直接译出。在此采用了翻译方法中的"转换法"，将其转译成条件状语从句，这样中文的译文更地道。

【2】The terminals of the source and load are interconnected by conductors (generally but not always wires).

该句中的 generally but not always wires =generally wires, but not always wires，译为"通常这种导体是导线，但少数情况下也有例外。"

【3】This simple circuit made up of a source, a load and two wires is seldom, if ever, met with in practice. =This simple circuit made up of a source, a load and two wires is seldom met with in practice, if it is ever met with in practice sometimes.

该句译为"这种由一个电源、一个负载和两根导线组成的简单电路，即使在实践中有时也能遇到，也是很少见的"。

【4】Whatever its origin (thermal, contact, etc.), the source e.m.f. E (Fig.2 (a), omitted) is numerically equal to the potential difference between terminals 1 and 2 with the external circuit open...

该句中的 between terminals 1 and 2 说明 difference，而 with the external circuit open 则修饰 terminals 1 and 2，因此此句译为"无论电动势 E 的原动力是什么（即不论是热的、机械的还是其他什么形式），其大小就等于 1 和 2 两端之间的开路电压"。

1.1.4　Translation

电路元件与参数

电路（或网络）是各电气装置的实物连接体。电路的作用是分配电能和转换能量形式。因此，电路的基本元件是电源、能量转换器以及它们之间的连接导线。

电源（如原生电池、再生电池和发电机等）将化学能、热能或其他形式的能量转换成电能。

能量转换器（也称为负载，比如灯泡、取暖器及电机等）将电能转换成光能、热能和机械能等。

电路的工作情况可以用电势（或电压）和电流来描述。当电能在产生、传输和变换时，若电路中相关的电流和电压不随时间而变化，我们便称其为直流电路。对于时不变的电流和电压，与电气设备相联系的电场和磁场也是时不变的。这也就是为什么在直流电路中没有自感和互感电动势以及在导体的周围电介质中也没有位移电流的原因。

图 1（略）用简化的方式表示以蓄电池为电源、以灯泡为负载的一个假想电路。电源和负载端由导体（通常是导线，但少数情况下也有例外）连接，如图所示，电源、负载和导体形成一个闭合回路。电动势产生一个绕该闭合回路的连续单向的电流。

这种由一个电源、一个负载和两根导线组成的简单电路，在实践中即使能遇到也是很少见的。实际电路可能包括许多按不同方式连接的电源和负载。

为简化对实际电路的分析，通常将它画成用符号表示的电路图，这种电路图实际上是虚构的，或更确切地说，是实际电路或网络的理想模型。这种电路由相互连接的电路元件或电路参数符号组成。为了表示一直流电路，至少应有两种元件，这就是电动势为 E、内阻为 R_s 的电源和负载电阻（包括连接导体的电阻）R（见图 2（略））。

无论电动势 E 的原动力是什么（即不论是热、摩擦还是其他形式产生的），电源电动势 E（见图 2（a）（略））在数值上等于 1 和 2 两端之间的开路电压，即当电源中无电流通过时

$$E = \varphi_1 - \varphi_2 = V_{12} \tag{1}$$

电源电动势的方向是从低电位点指向高电位点，在电路图中用箭头表示。

当一负载与电源相连（此时电路称为已载荷）形成闭合回路时，在此回路中便有电流。由于电源内部的压降 V_s（也即内阻 R_s 上的压降）

$$V_s = R_s I$$

这时电源 1 和 2 两端之间（也称端电压）便不等于它的电动势。

图 3（略）表示了一个带负载后电源的典型的所谓外特性 $V_{12} = \varphi_1 - \varphi_2 = VI$（也称为电源的负载特性）。从图中可看出，当电流从零增大到 $I \approx I_1$ 时，电源端电压将线性下降

$$V_{12} = V = E - V_s = E - R_s I$$

换句话说，电源内阻两端的压降 V_s 与电流成正比，该过程一直持续到电流达到某一临界值为止。然后，随着电流的继续增大，其与电源端电压之比值便发生了变化，外特性也不再为线性的。电压的这种下降也许是由于电源电压的下降，也许是由于内阻的增大，或者两者兼而有之。

电源提供的功率由下式确定

$$P_s = EI \tag{2}$$

式中，P_s 为电源功率。

在此看来应该消除关于功率的一种通常错误的概念，比如人们可能听说过关于功率的产生、提供、消耗、传输、损耗等说法。然而，事实上，只有能量才有产生、提供、消耗、传输和损耗的说法，功率仅是能量的输入或转换的比率，即单位时间内产生、提供和传输等的能量值。因此，在上述内容中用"能量"这个术语而不用"功率"会更准确些，不过人们习惯了传统的说法。

作为一种抽象化的电路元件，负载电阻 R（见图 2（b）（略））形成了一个消耗能量的概念，即将电能转换成热量，因此定义为

$$P = R I^2 \tag{3}$$

通常，负载电阻仅取决于通过负载的电流，这一点可用函数符号 $R(I)$ 来表示。由欧姆定律可知，电阻两端的电压为

$$V=RI \quad (4)$$

在电路分析中，常常使用电阻的倒数，称之为电导，其定义为

$$g=1/R$$

在实际问题中，人们并不常常将电阻表示为电流的函数 $R(I)$，而将电阻两端的电压表示为电流的函数 $V(I)$，或其反函数 $I(V)$。函数 $V(I)$ 或 $I(V)$ 的关系已成为人所共知的伏安特性。

图4（略）中的曲线 $V_1(I)$ 和 $V_2(I)$ 分别表示一金属丝灯泡和碳丝灯泡的伏安特性曲线。如图所示每个灯泡的电流和电压之间的关系并不是线性的。随着电流的增加，金属丝灯泡的电阻是增加的，而碳丝灯泡的电阻是减小的。

含有非线性元件的电路称为非线性电路。

假如电源的电动势和内阻与其连接的负载电阻被认为均不随电流和电压变化而变化，那么电源的外特性 $V(I)$ 和负载的伏安特性 $V_1(I)$ 将为线性的（见图5（略））。

仅含线性元件的电路称为线性电路。

大多数实际电路可归类为线性电路。因此，对线性电路性能和线性电路分析的研究就具有理论和实践的双重意义。

Unit 1.2 Ideal Sources Series and Parallel Equivalent Circuits

1.2.1 Text

Consider an elementary circuit containing a single source of e.m.f. E and of internal resistance R_s, and a single load R (Fig.1, omitted). The resistance of the conductors of this type of circuit may be neglected. In the external portion of the circuit, that is, in the load R, the current is assumed to flow from the junction a (which is at a higher potential such that $\varphi_a = \varphi_1$) to the junction b (which is at a lower potential such that $\varphi_b = \varphi_2$). The direction of current flow may be shown either by a hollow arrowhead or by supplying the current symbol with a double subscript whose first digit identifies the junction at a higher potential and the second the junction at a lower potential[1]. Thus for the circuit of Fig.1 (omitted), the current $I = I_{ab}$.

We shall show that the circuit of Fig.1 (omitted) containing a source of known e.m.f. E and source resistance R may be represented by two types of equivalent circuits.

As already started, the terminal voltage of a loaded source is lower than the source e.m.f. by an amount equal to the voltage drop across the source resistance:

$$V = \varphi_1 - \varphi_2 = E - V_s = E - R_s I \tag{1}$$

On the other hand, the voltage across the load resistance R is

$$V = \varphi_a - \varphi_b = RI \tag{2}$$

Since $\varphi_1 = \varphi_a$ and $\varphi_2 = \varphi_b$, from Eqs. (1) and (2), it follows that

$$E - R_s I = RI$$

or

$$E = R_s I + RI \tag{3}$$

and

$$I = E/(R_s + R)$$

From the last equation we conclude that the current through the source is controlled by both the load resistance and the source resistance. Therefore, in an equivalent circuit diagram the source resistance R_s may be shown connected in series with the load resistance R. This configuration may be called the series equivalent circuit (usually known as the Thevenin equivalent source).

Depending on the relative magnitude of the voltages across R_s and R, we can develop two modifications of the series equivalent circuit. In the equivalent circuit of Fig.2 (a) (omitted), V is controlled by the load current and is decided by the difference between the source e.m.f. E and the voltage drop V_s. If $R_s \ll R$ and, for the same current, $V_s \ll V$ (that is, if the source is operating under conditions very close to no-load or an open-circuit), we may neglect the internal voltage drop, put $V_s = R_s I = 0$ (very nearly) and obtain the equivalent circuit of Fig.2 (b) (omitted). What we have got is a source whose internal resistance is zero ($R = 0$). It is called an ideal voltage source. In diagrams it is symbolized by a circle with an arrow inside and the letter E beside it. When applied to a network, it is called a driving force or an impressed voltage source. The terminal voltage of an ideal voltage source is independent of the load resistance and is always equal to the e.m.f. E of the practical source it represents. Its external characteristic is a straight line parallel to the x-axis (the dotted line ab in Fig.3). The other equivalent circuit (Fig.3, omitted) may be called the parallel equivalent circuit (usually known as the Norton equivalent). It may also have two modifications. To prove this, we divide the right- and left-hand sides of Eq. (3) by R_s:

$$E/R_s = I + V/R_s = I + V_{gs}$$

or
$$J = I + I_s \qquad (4)$$

where $J = E/R_s$ current with the source short-circuited (with $R=0$); $I_s = V/R_s = V_{gs}$ current equal to the ratio of the terminal source voltage to the source resistance; $I = V/R$ load current.

Eq. (4) is satisfied by the equivalent circuit of Fig.3 (a) (omitted), in which the source resistance R_s is placed in parallel with the load resistance R.

If $g_s \ll g$ or $R_s \gg R$ and, for the same voltages across R_s and R, the current $I_s \ll I$ (that is, if the source is operating under conditions approaching a short-circuit), we may put $I_s = V_{gs} = 0$ (very nearly) and get the equivalent circuit of Fig.3 (b) (omitted).

What we have got is a source of zero internal conductance, $g_s = 0$ ($R_s = \infty$). It is called an ideal current source. When applied to a circuit, it is called a driving force of an impressed current source. The current of an ideal current source is independent of the load resistance R and is equal to E/R_s. The external characteristic of an ideal current source is a straight line parallelled to the y-axis (the dotted line in Fig.3 (omitted)).

Thus whether a real energy source may be represented by an ideal voltage source or an ideal current source depends on the relative magnitude of R_s and R. A real source, though, may be represented by an ideal voltage or current source also when R_s is comparable with R. In such a case, either R_s, or $g_s = 1/R_s$ (Figs.2 (a) and 3 (a), respectively) should be removed from the source and lumped with R or $g = 1/R$.

Ideal voltage and current sources are active circuit elements, while resistances and conductance are passive elements.

In developing an equivalent circuit, it is important to take into account, as much as practicable, the known properties of each device and of the circuit as a whole.

Let us develop an equivalent circuit for a two-wire power transmission line of length l, diagrammatically shown in Fig.4 (a) (omitted). There is a generator of e.m.f. E and of source resistance R_s at the sending end and a load of resistance R_2 at the receiving end of the line.

It is obvious that the receiving end voltage will be less than the sending end one by an amount equal to the voltage drop across the resistance of the line conductors. The current at the receiving end will be smaller than that at the sending end by an amount equal to the leakage current (due to imperfect insulation).

Let each line conductor have a resistance $R_0/2$ and a conductance g_0 per unit length of the line. We divide the line into length elements dx (Fig.4 (a), omitted). Then each length element will have associated with it the combined resistance of the "go" and "return" wires, $R_0\,\mathrm{d}x = (R_0/2)\,\mathrm{d}x + (R_0/2)\,\mathrm{d}x$ and a conductance $g_0\,\mathrm{d}x$【2】. Accordingly, the entire line may be represented by a network of elements each of resistance $R_0\,\mathrm{d}x$ and conductance $g_0\,\mathrm{d}x$ (Fig.4 (b), omitted). The sending end generator in this network is represented by a voltage source (of e.m.f. E and of source resistance R_s).

From this equivalent circuit we can find the voltage and current at any point on the line in terms of specified voltage and current at the sending or receiving end.

If the leakage current of the line is only a small fraction of the load current, we may neglect it and remove all conductance $g_0\,\mathrm{d}x$ from the network. This will leave us with a simple, single-loop network with one and the same current in each of its elements, such as shown in Fig.5 (omitted), where the line resistance $R_\mathrm{line} = R_0\,l$ is connected in series with R_s and R_2. The network of Fig.5 (omitted) may be used for analysis of line performance without a consideration of leakage currents.

It takes some practice to learn to develop equivalent circuits such as will reflect the behavior of physical prototypes in the most faithful manner and meet the requirements of the problem at hand【3】.

1.2.2　Specialized English Words

ideal source　理想电源
series and parallel equivalent circuit　串并联等值电路
internal resistance　内阻
sending end　发送端
double subscript　双下标
ideal voltage source　理想电压源
ideal current source　理想电流源
active circuit elements　有源电路元件
passive circuit elements　无源电路元件
power transmission line　输电线
receiving end　接收端
leakage current　漏电流

1.2.3　Notes

【1】The direction of current flow may be show either by a hollow arrowhead or by supplying the current symbol with a double subscript whose first digit identifies the junction at a higher potential and the second the junction at a lower potential.

此句是一个主从复合句，主语是 The direction of...a double subscript，意为"电流的方向既可以用一个空心箭头来表示，也可以用带有双下标的电流符号来表示"，只是为了定义这个双下标符号，用了一个由 whose 引导的定语从句来修饰 a double subscript，这个从句从 whose 起一直到句末，其本身实际上是一个并列句，前一个子句是 first digit identifies the junction at a higher potential，其中 first digit 是主语，identifies 是谓语，at a higher potential 是介词短语做 the junction 的后置定语，本子句的意思是"第一个数字表示相对高电位的那一点"。后面那一部分实际上是另一个子句的省略部分，即 the second the junction at a lower potential=the second digit identifies the junction at a lower potential，因为其句子中的谓语动词在前一子句中已出现，且此两个子句在形式上完全相同，故将其谓语动词省略。

【2】Then each length element will have associated with it the combined resistance of the "go" and "return" wires，$R_0 \mathrm{d}x = (R_0/2)\mathrm{d}x + (R_0/2)\mathrm{d}x$ and a conductance $g_0 \mathrm{d}x$.

此句应该翻译成"这样每一长度元将有与之相关的'去'与'来'的组合电阻 $R_0 \mathrm{d}x = (R_0/2)\mathrm{d}x + (R_0/2)\mathrm{d}x$ 和电导 $g_0\mathrm{d}x$"。

【3】It takes some practice to learn to develop equivalent circuits such as will reflect the behavior of physical prototypes in the most faithful manner and meet the requirements of the problem at hand.

此句的主句是 It takes some practice to learn to develop equivalent circuits，其中 to learn to develop equivalent circuits 是不定式短语做主语，只是由于其太长，若放在句首，则会给人头重脚轻的感觉，所以用 It 做形式主语。其次要了解此句是 It takes something（time，money and so on）…的句型，此处是 practice，其意思是"（这件事）是要花一些时间，要花一些金钱，要花一番工夫的"。这样，主句的意思就清楚了，"学会画（这样的）等值电路是要花一番工夫的"。那么是什么样的等值电路呢？后面整个都是由 such as（that）引导的定语从句，修饰 equivalent circuits，意为"既能以最可靠的方式反映实际电路的运行情况，又能满足常常遇到的实际问题的需要的这样一种等值电路"。

1.2.4 Translation

理想电源的串并联等值电路

假设有一个电动势为 E、内阻为 R_s 的电源和一个负载 R 构成的基本电路（如图 1（略）所示），并且这种电路的连接导体的电阻可以忽略不计。在电路的外部，也即在负载 R 中，假定电流从端点 a（位于 $\varphi_a = \varphi_1$ 的高电位）流向端点 b（位于 $\varphi_b = \varphi_2$ 的低电位）。电流的方向既可用一个空心箭头来表示，也可以用带有双下标的电流符号来表示，且第一个下标认定为高电位点，第二个下标为低电位点，于是，图 1 的电流 $I=I_{ab}$。

下面将证明，含一个已知电动势 E 和内阻 R，电源（如图 1（略）所示）电路可用两种类型的等效电路来表示。

如前所述，电源加负载后的端电压比电源电动势小一个电源内阻上的压降

$$V = \varphi_1 - \varphi_2 = E - V_s = E - R_s I \tag{1}$$

另外，负载电阻 R 的电压为

$$V = \varphi_a - \varphi_b = RI \tag{2}$$

由于 $\varphi_1 = \varphi_a$ 和 $\varphi_2 = \varphi_b$，从式（1）和式（2）可得

$$E - R_s I = RI$$

或

$$E = R_s I + RI \tag{3}$$

和

$$I = E/(R_s + R)$$

从上式我们可得出，流过电源的电流是由负载电阻和电源内阻二者所决定的。于是，在等效电路图中，电源电阻 R_s 可表示成与负载电阻 R 相串联。这种连接方式称为串联等效电路（也就是人们所熟知的戴维南等效电源）。

根据 R_s 和 R 两端电压的相对大小，我们可推出两种串联电路的变换形式。在图 2（a）（略）等效电路中，V 由负载电流所控制，并由电源电动势 E 和电压降 V_s 的差值所决定。假如 $R_s \ll R$，对同一电流，$V_s \ll V$（也就是电源近乎工作在无载，即开路状态），这样可忽略内压降，令 $V_s = R_s I = 0$（非常近似），因而得到如图 2（b）（略）所示的等效电路。这是一个内阻为零的电源，称之为理想电源，在电路图中它是用一个里面有一个箭头、旁边标有字母 E 的圆圈表示。当接入电路时称之为驱动势或电压源。理想电压源的端电压与负载电阻无关，它恒等于其所表示的实际电源电动势 E。它的外特性是一条平行于 x 坐标轴的直线（如图 3（略）所示的虚线 ab）。另一个等效电路可称为并联等效电路（也就是人们所熟知的诺顿等效电路）。等效电路可有两种变换形式，为了证明这一点，我们可用 R_s 同时除式（3）的左边和右边

$$E/R_s = I + V/R_s = I + V_{gs}$$

或

$$J = I + I_s$$

式中，$J = E/R_s$ 为电源短路时的电流（即 $R=0$ 时）；$I_s = V/R_s = V_{gs}$ 为电源端电压与电源内阻的比值所确定的电流；$I = V/R$ 为负载电流。

式（4）与图 3（a）（略）所示的等效电路相符，在该图中，电源内阻 R_s 与负载电阻 R 相并联。假如 $g_s \ll g$ 或 $R_s \gg R$ 且 R_s 和 R 上的电压相同，因此电流 $I_s \ll I$（也就是电源近乎工作在短路状态），因此可以令 $I_s = V_{gs} = 0$（非常近似），这样便得如图 3（b）（略）所示的等效电路。

这是一个内部电导 $g_s = 0$（$R_s = \infty$）的电源。理想电流源的电流等于 E/R_s，与负载电阻 R 无关。理想电流源的外特性是一条平行于坐标轴 y 的直线（如图 3（略）虚线 cd 所示）。因此，一个实际的电源表达成理想电压源还是理想电流源，由 R_s 和 R 的相对大小所决定。

然而，当一个实际电源的 R_s 与 R 可相比较时，既可用理想电压源表示，又可用理想电流源表示。在这种情况下，应将 R_s 或 $g_s = 1/R_s$（如图 2（a）（略）和图 3（a）（略）所示）从电源中移出并与 R 或 $g = 1/R$ 进行合并。

理想电压源和理想电流源是有源的电路元件，而电阻和电导是无源的电路元件。

在推导等效电路时，重要的是尽可能把每一装置和整个电路的已知特性都考虑到。

下面推导如图 4（a）（略）所示的长度为 l 的两根输电线的等效电路。该网络中，在发送端有电动势为 E、内阻为 R_s 的电源，在接收端有一个负载电阻 R_2。

很明显，接收端的电压比发出端的电压小一个导线上的压降，接收端的电流比发出端的电流小一个漏电流（由于不完全绝缘）。

令每一根导体单位长度的电阻为 $R_0/2$，单位长度的电导为 g_0。

我们将输电线分成长度元 dx（如图 4（a）（略）所示），这样每一长度元将有与之相关的"去"与"来"的组合电阻 $R_0 dx = (R_0/2)dx + (R_0/2)dx$ 和电导 $g_0 dx$。因此，整条输电线可用单元组成的网络表示，每一单元含有 $R_0 dx$ 和 $g_0 dx$，如图 4（b）（略）所示。该网络中的发送端的发电机用一个电压源来表示（电动势为 E，电源内阻为 R_s）。

从这个等效电路我们能求出以发送端或接收端的电压或电流的形式表示的，输电线上任一点的电流和电压。

若输电线的漏电流只是负载电流很小的一部分，便可以将它忽略，并把电导 $g_0 dx$ 从电路

中移去。这样剩下的是一个简单的、单回路的、通过各元件的电流是同一个电流的电路，如图 5（略）所示。图中线电阻 $R_{\text{line}}=R_0 l$ 与 R_s、R_2 是串联的。图 5 所示的电路可用于分析不涉及漏电流的输电线路。

要学会建立既能以最可靠的方式反映实际电路的运行情况，又能满足解决常常遇到的实际问题需要的等值电路是要下一番工夫的。

Unit 1.3　Analysis of Sinusoidal Alternating Electricity

1.3.1　Text

R.M.S.（Effective）Values of Current and Voltage

The force between two current-carrying conductors is proportional to the square of the current in the conductors. The heat due to a current in a resistance over a period is also proportional to the square of that current. This calls for knowledge of what is known as the root mean square（or effective）current defined as

$$I = \sqrt{\left(\frac{1}{T}\int_0^T i^2 dt\right)} \tag{1}$$

The heat developed by a current i in a resistance r in time dt is

$$\int_0^T ri^2 dt = rT\frac{1}{T}\int_0^T i^2 dt = rI^2 T$$

It follows that the r.m.s.（effective）value of an alternating current is numerically equal to the magnitude of the steady direct current that would produce the same heating effect in the same resistance and over the same period of time 【1】.

Let us establish the relationship between the r.m.s. and peak values of a sinusoidal current, I and I_m

$$I^2 = \frac{1}{T}\int_0^T i^2 dt = \frac{I_m^2}{T}\int_0^T \sin^2(\omega t + \varphi)dt = \frac{I_m^2}{2T}\int_0^T [1 - \cos(2\omega t + 2\varphi)]dt = I_m^2/2$$

Hence
$$I = I_m/\sqrt{2} \tag{2}$$

The r.m.s.（effective）values of e.m.f. and voltage are

$$E = \sqrt{\left(\frac{1}{T}\int_0^T e^2 dt\right)} = E_m/\sqrt{2}$$

and
$$V = \sqrt{\left(\frac{1}{T}\int_0^T v^2 dt\right)} = V_m/\sqrt{2}$$

In dealing with periodic voltages and currents, their r.m.s.（effective）values are usually meant, and the adjective "r.m.s." or "effective" is simply implied 【2】.

Representation of Sinusoidal Time Functions by Vectors and Complex Number

A.C. circuit analysis can be greatly simplified if the sinusoidal quantities involved are represented by vectors or complex numbers.

Let there be a sinusoidal time function（current, voltage, magnetic flux and the like）：
$$v = V_m \sin(\omega t + \varphi)$$

It can be represented in vector form as follows. Using a right-hand set of Cartesian coordinates MON（Fig.1, omitted）, we draw the vector $\dot{V}_m$ to some convenient scale such that it represents the peak value V_m and makes the angle φ with the horizontal axis OM (positive values of φ are laid off counter-clockwise, and negative, clockwise). Now we imagine that, starting at $t=0$, the vector $\dot{U}_m$ begins to rotate about the origin O counter-clockwise at a constant angular velocity equal to the angular frequency ω. At time t, the vector makes the angle $\omega t + \varphi$ with the axis OM. Its projection onto the vertical axis $N'N$ represents the instantaneous value of v to the scale chose.

Instantaneous values of v, as projections of the vector on the vertical axis $N'N$, can also be obtained by holding the vector $\dot{V}_m$ stationary and rotating the axis $N'N$ clockwise at the angular velocity ω, starting at time $t=0$. Now the rotating axis $N'N$ is called the time axis. In each case, there is a single-valued relationship between the instantaneous value of v and the vector $\dot{V}_m$. Hence $\dot{V}_m$ may be termed the vector of the sinusoidal time function v. likewise, there are vectors of voltages, e.m.f.s, currents, magnetic fluxes, etc.

"True" vector quantities are demoted either by clarendon type, e.g. $\boldsymbol{A}$, or by $\overline{A}$, while sinusoidal ones are demoted by $\dot{A}$. Graphs of sinusoidal vectors, arranged in a proper relationship and to some convenient scale, are called vector diagrams.

Taking MM' and NN' as the axes of real and imaginary quantities, respectively, in a complex plane, the vector $\dot{V}_m$ can be represented by a complex number whose absolute value (or modulus) is equal to V_m, and whose phase (or argument) is equal to the angle φ 【3】. This complex number is called the complex peak value of a given sinusoidal quantity.

Generally, a complex vector may be expressed in the following ways:

$$\left.\begin{array}{l} \dot{V} = V_m \angle \varphi \quad (ploar) \\ \dot{V} = V_m e^{j\varphi} \quad (exponential) \\ \dot{V} = V_m(\cos\varphi + j\sin\varphi) \quad (trigonometric) \\ \dot{V} = V'_m + jV''_m \quad (tectangular\ of\ algebraic) \end{array}\right\} \quad (3)$$

where $j=\sqrt{-1}$.

When the vector $\dot{V}_m$ rotates counter-clockwise at angular velocity ω, starting at $t=0$, it is said to be a complex time function, defined so that $\dot{V}_m = V_m e^{j(\omega t + \varphi)}$. Now, since this is a complex function it can be expressed in terms of its real and imaginary parts:

$$V_m e^{j(\omega t + \varphi)} = V_m \cos(\omega t + \varphi) + jV_m \sin(\omega t + \varphi)$$

where the sine term is the imaginary part of the complex variable equal (less j) to the sinusoidal quantity v, or

$$v = \text{Im}[e^{j(\omega t + \varphi)}] = \text{Im}[V_m e^{j\varphi} e^{j\omega t}] = \text{Im}[\dot{V}_m e^{j\omega t}] \quad (4)$$

where the symbol Im indicates that only the imaginary part of the function in the square brackets is taken.

The instantaneous value of a cosinusoidal function is given by

$$v = V_m \cos(\omega t + \theta) = \text{Re}[V_m \cos(\omega t + \theta) + jV_m \sin(\omega t + \theta)]$$
$$= \text{Re}[V_m e^{j(\omega t + \theta)}] = \text{Re}[\dot{V}_m e^{j\omega t}] \quad (5)$$

where the symbol Re indicates that the real part of the complex variable in the square brackets is only taken. For this case, the instantaneous value of v is represented by a projection of the vector $\dot{V}_m e^{j\omega t}$ onto the real axis. The representation of sinusoidal functions in complex form is the basis of the complex-number method of A.C. circuit analysis. In its present form, the method of complex numbers was introduced by Heaviside and Steinmetz.

Addition of Sinusoidal Time Functions

A.C. circuit analysis involves the addition of harmonic time functions having the same frequencies but different peak values and epoch angles. Direct addition of such functions would call for unwieldy trigonometric transformations. Simpler approaches are provided by the Argand diagram (graphical solution) and by the method of complex numbers (analytical solution).

Suppose we are to find the sum of two harmonic functions:
$$v_1 = V_{m1} \sin(\omega t + \varphi_1)$$
and
$$v_2 = V_{m2} \sin(\omega t + \varphi_2)$$

First, consider the application of the Argand diagram (graphical solution). We lay off the vectors $\dot{V}_{m1} = V_{m1} \angle \varphi_1$ and $\dot{V}_{m2} = V_{m2} \angle \varphi_2$ and find the resultant vector $\dot{V}_m = V_m \angle \varphi$.

Now assume that the vectors $\dot{V}_{m1}$, $\dot{V}_{m2}$, $\dot{V}_m$ begin to rotate about the origin of coordinates, O, at $t=0$, doing so with a constant angular velocity ω in the counter-clockwise direction.

At any instant of time, a projection of the rotating vector $V_m \angle (\omega t + \varphi)$ onto the vertical axis $N'N$ is equal to the sum of projections onto the same axis of the rotating vectors $V_{m1} \angle (\omega t + \varphi_1)$ and $V_{m2} \angle (\omega t + \varphi_2)$, or the instantaneous values v_1 and v_2. In other words, the projection of $V_m \angle (\omega t + \varphi)$ onto the vertical axis represents the sum $(v_1 + v_2)$, and the vector $\dot{V}_m = V_m \angle (\omega t + \varphi)$ represents the desired sinusoidal time function $v = v_1 + v_2$.

On finding the length of V_m and the angle φ from the Argand diagram, we may substitute them in the expression $v = V_m \sin(\omega t + \varphi)$.

Now consider the analytical method. Referring to the diagram of Fig.2 (omitted), we may write
$$\dot{V}_{m1} + \dot{V}_{m2} = \dot{V}_m$$

In the rectangular (algebraic) form, these complex numbers are $\dot{V}_{m1} = V'_{m1} + jV''_{m1}$, $\dot{V}_{m2} = V'_{m2} + jV''_{m2}$.

On adding them together we obtain
$$V'_{m1} + jV''_{m1} + V'_{m2} + jV''_{m2} = V'_m + jV''_m = \dot{V}_m$$

where $V_m = \sqrt{(V'_m)^2 + (V''_m)^2}$, $\tan \varphi = V''_m / V'_m$.

Since $\tan \varphi = \tan(\varphi \pm \pi)$, it is important to know the quadrant where $\dot{V}_m$ occurs, before we can determine φ. The quadrant can be readily identified by the signs of the real and imaginary parts of the function. For convenience the epoch angle φ may be expressed in degrees rather than in radians.

The two methods are applicable to the addition of any number of sinusoidal functions of the same frequency.

In practical work, one is usually interested in the r.m.s. values and phase displacements of sinusoidal quantities. Therefore the Argand diagram is simplified by omitting the axes (whose position is immaterial), while the phase displacement between the vectors is faithfully reproduced. Also, instead of rotating vectors of length equal to the peak values of sinusoidal quantities, the scale is changed and the vector lengths are treated as r.m.s. values.

In analytical treatment, it is usual to arrange the sinusoidal quantities so that the epoch angle of any one becomes zero. Likewise, instead of complex peak values, the respective complex r.m.s. values, obtained by division of complex peak values by $\sqrt{2}$, are used. For brevity, complex r.m.s. values are called simply a complex current, a complex voltage, etc.

1.3.2　Specialized English Words

r.m.s. values= root of mean square　均方根值
effective values　有效值
steady direct current　恒稳直流电
sinusoidal time function　正弦时间函数
vector　矢量
complex number　复数
Cartesian coordinates　笛卡儿坐标系
counter-clockwise　逆时针方向
clockwise　（顺）时针方向
trigonometric transformations　三角转换
Argand　阿尔冈，法国数学家
absolute value　绝对值
modulus　（复数）模
complex peak value　复数的幅值
complex time function　复数时间函数
real part　（复数）实部
imaginary part　（复数）虚部
complex-number method=method of complex numbers　复数法
Heaviside　赫维赛德，美国物理学家
Steinmetz　施太因梅兹，出生于法国的美国电气工程师，提出交流电系统概念，创立计算交流电路的方法，研究电瞬变现象理论，著有《交流电现象的理论和计算》
vector diagrams　矢量图
complex plane　复平面
vectors of voltages（currents，magnetic fluxes，etc.）　电压（电流、磁通等）矢量
analytical solution　解析法
Argand diagram　阿尔冈图
origin of coordinates　坐标原点
rotating vector　旋转矢量
instantaneous values　瞬时值
epoch angle　初相角
phase displacement　相位差
constant angular velocity　恒定角速度

1.3.3 Notes

【1】It follows that the r.m.s.(effective) value of an alternating current is numerically equal to the magnitude of the steady direct current that would produce the same heating effect in the same resistance and over the same period of time.

句中 It follows that…是主句，that 后面是其引导的宾语从句，此句型固定译为"由此得出……"。只是该宾语从句中又有一个修饰 the steady direct current 的定语从句。that would produce the same heating effect in the same resistance and over the same period of time，意为"在相同的电阻上和同一时间内产生相同热效应的"，因此，整句译为"由此得出，交流电流的均方根（有效）值在大小上等于在相同的电阻上和同一时间内产生相同的热效应的恒稳直流电流"。

【2】In dealing with periodic voltages and currents, their r.m.s.(effective) values are usually meant, and the adjective "r.m.s." or "effective" is simply implied.

该句中，第一个并列句直译为"均方根（有效）值被通常所指"，因此可转译为"通常指的就是均方根（有效）值"；第二句直译为"限定词'均方根'或'有效的'只是暗指（隐意）"，因此可转译为"便将限定词'均方根'或'有效的'略去，并不指明"，这就是翻译方法中通常所说的"意译"。

【3】Taking MM' and NN' as the axes of real and imaginary quantities, respectively, in a complex plane, the vector $\dot{V}_m$ can be represented by a complex number whose absolute value (or modulus) is equal to V_m, and whose phase (or argument) is equal to the angle φ.

分析这种长句，应学会先将句干找出来，而将其他成分暂时搁置不看，特别是一些用逗号分开的短语，可以将其当作插入语看待，暂时不予理会。这样便是，Taking MM' and NN' as the axes of real and imaginary quantities…, the vector $\dot{V}_m$ can be represented by a complex number…，可见整个句子的框架结构是带有一现在分词短语做状语的句子结构，意为"取 MM' 和 NN' 作为实轴和虚轴的量，矢量 $\dot{V}_m$ 便可由一复数表示。"只是在现在分词短语中修饰具有动词行为的分词 Taking 还有两个状语，一个是方式状语 respectively，一个是地点状语 in a complex plane，只是因为状语太多，便用逗号分开。弄清这种关系后，再将其表达的意思加到分词短语中去就行了，即整个分词短语可译成"在一复平面上，（若）取 MM' 和 NN' 作为实轴和虚轴的量，便……"。同理，主句也是如此，在 a complex number 后有两个由 whose 引导出的定语从句 whose absolute value (or modulus) is equal to V_m, and whose phase (or argument) is equal to the angle φ. 意为"（这个复数）的绝对值（也即模）等于 V_m，它的相位（也即相角）等于角 φ"。

1.3.4 Translation

正弦交流电的分析

电流和电压的有效值

两载流导体之间的作用力与导体中的电流的平方成正比。某段时间内电流通过一个电阻所产生的热量也正比于电流的平方。这便引出通常所说的均方根（或有效值）电流的概念，其定义如下

$$I = \sqrt{\frac{1}{T}\int_0^T i^2 \mathrm{d}t} \tag{1}$$

在 $\mathrm{d}t$ 时间里电流 i 通过电阻 r 产生的热量为

$$\int_0^T ri^2 \mathrm{d}t = rT \cdot \frac{1}{T}\int_0^T i^2 \mathrm{d}t = rI^2 T$$

由此可得出，交流电流的均方根（或有效值）在数值上等于在相同电阻、相同时间内产生相同热量的直流电的电流。

下面求解正弦电流有效值和幅值之间，即 I 与 I_m 的关系

$$I^2 = \frac{1}{T}\int_0^T i^2 \mathrm{d}t = \frac{I_m^2}{T}\int_0^T \sin^2(\omega t + \varphi)\mathrm{d}t$$

$$= \frac{I_m^2}{2T}\int_0^T [1 - \cos(2\omega t + 2\varphi)]\mathrm{d}t = I_m^2 / 2$$

于是

$$I = I_m / \sqrt{2} \tag{2}$$

电动势和电压的有效值为

$$E = \sqrt{\frac{1}{T}\int_0^T e^2 \mathrm{d}t} = E_m / \sqrt{2}$$

和

$$V = \sqrt{\frac{1}{T}\int_0^T v^2 \mathrm{d}t} = V_m / \sqrt{2}$$

在涉及交流电压和电流时，通常指的值就是其均方根（有效）值，便将限定词"均方根（有效）"几个字略去，并不明指。

正弦时间函数的矢量和复数表示法

如果所涉及的正弦量用矢量和复数表示，便可大大地简化。

设一正弦时间函数（电流、电压、磁通等）

$$v = V_m \sin(\omega t + \varphi)$$

它可用如下的矢量形式表示。在笛卡儿坐标系的右侧 MON（如图1（略）所示）区域内取恰当的比例画出矢量 $\dot{V}_m$，以便于代表该量的幅值，并与横坐标形成 φ 角（逆时针方向为正，顺时针方向为负）。现在假设从 $t=0$ 开始，矢量 $\dot{V}_m$ 绕着原点 O 以角频率 ω 的恒定角速度逆时针旋转，t 时刻矢量与横坐标轴 OM 形成 $\omega t + \varphi$ 的夹角。它在纵坐标轴 $N'N$ 上的投影便表示在已选用的比例尺下的瞬时值 v。

瞬时值 v（即矢量在纵坐标轴 $N'N$ 上的投影）也能通过以下方法得到，即令矢量 $\dot{V}_m$ 不动，将轴 $N'N$ 以角速度 ω 从 $t=0$ 开始顺时针旋转，此时旋转的轴 $N'N$ 称为时间轴。

两种情况下，瞬时值 v 和矢量 $\dot{V}_m$ 之间都存在单值关系。因此，$\dot{V}_m$ 便可称为正弦时间函数 v 的矢量。同理，还有电压矢量、电势矢量、电流矢量、磁通矢量等。

真正的矢量是用粗体字 $\mathbf{A}$，或 $\overline{A}$ 表示的，而正弦矢量则用 $\dot{A}$ 表示。按合适的相对关系和某种方便的比例画出的正弦相量的图解称为矢量图。

在一复数平面内，取 MM' 和 NN' 分别为实数轴和虚数轴，矢量 $\dot{V}_m$ 可用一复数来表示，该复数的绝对值（即模）等于 V_m，其相位角等于 φ。此复数称为某一已知正弦量的最大值复数。

通常，复数可用复矢量表示为如下形式：

$$\left.\begin{array}{ll}\dot{V}=V_{\mathrm{m}}\angle\varphi & \text{（极坐标的）}\\ \dot{V}=V_{\mathrm{m}}\mathrm{e}^{\mathrm{j}\varphi} & \text{（指数的）}\\ \dot{V}=V_{\mathrm{m}}(\cos\varphi+\mathrm{j}\sin\varphi) & \text{（三角的）}\\ \dot{V}=V'_{\mathrm{m}}+\mathrm{j}V''_{\mathrm{m}} & \text{（直角或代数的）}\end{array}\right\} \quad (3)$$

式中，$j=\sqrt{-1}$。

当矢量 $\dot{V}_{\mathrm{m}}$ 从 $t=0$ 开始以角速度 ω 逆时针旋转时，便被称之为复数时间函数，并定义为 $\dot{V}_{\mathrm{m}}=V_{\mathrm{m}}\mathrm{e}^{\mathrm{j}(\omega t+\varphi)}$。现在，既然它是一复函数，因此便可用实部和虚部来表示

$$V_{\mathrm{m}}\mathrm{e}^{\mathrm{j}(\omega t+\varphi)}=V_{\mathrm{m}}\cos(\omega t+\varphi)+\mathrm{j}V_{\mathrm{m}}\sin(\omega t+\varphi)$$

式中正弦项是复数变量（除去 j）的虚部，等于正弦量 v，即

$$\begin{aligned}v&=\mathrm{Im}[V_{\mathrm{m}}\mathrm{e}^{\mathrm{j}(\omega t+\varphi)}]=\mathrm{Im}[V_{\mathrm{m}}\mathrm{e}^{\mathrm{j}\varphi}\mathrm{e}^{\mathrm{j}\omega t}]\\ &=\mathrm{Im}[\dot{V}_{\mathrm{m}}\mathrm{e}^{\mathrm{j}\omega t}]\end{aligned} \quad (4)$$

式中符号 Im 是指只计及方括号中复数的虚部。

余弦函数的瞬时值由下式给出

$$\begin{aligned}v&=V_{\mathrm{m}}\cos(\omega t+\theta)\\ &=\mathrm{Re}[V_{\mathrm{m}}\cos(\omega t+\theta)+\mathrm{j}V_{\mathrm{m}}\sin(\omega t+\theta)]\\ &=\mathrm{Re}[V_{\mathrm{m}}\mathrm{e}^{\mathrm{j}(\omega t+\theta)}]\\ &=\mathrm{Re}[\dot{V}_{\mathrm{m}}\mathrm{e}^{\mathrm{j}\omega t}]\end{aligned} \quad (5)$$

式中符号 Re 是指只计及方括号中复数的实部。在这种情况下，瞬时值由矢量 $\dot{V}_{\mathrm{m}}\mathrm{e}^{\mathrm{j}\omega t}$ 在实轴上的投影表示，复数形式的正弦函数表达式是交流电路分析中复数法的基础。现在所用的复数符号法的形式是由赫维赛德和施泰因梅兹提出的。

正弦时间函数的加法

交流电路的分析包括对有相同频率、不同幅值和初相角的谐振时间函数的加法。这些函数直接相加将要求用到繁杂的三角转换。简单的方法是采用阿尔冈图法（图解法）和复数方法（解析方法）。

假如要求两个谐振函数的和

$$v_1=V_{\mathrm{m}1}\sin(\omega t+\varphi_1)$$

和

$$v_2=V_{\mathrm{m}2}\sin(\omega t+\varphi_2)$$

首先，考虑采用阿尔冈图法（作图法）。画出矢量 $\dot{V}_{\mathrm{m}1}=V_{\mathrm{m}1}\angle\varphi_1$ 和 $\dot{V}_{\mathrm{m}2}=V_{\mathrm{m}2}\angle\varphi_2$，并由平行四边形法则求出合成矢量 $\dot{V}_{\mathrm{m}}=V_{\mathrm{m}}\angle\varphi$。

现在假定矢量 $\dot{V}_{\mathrm{m}1}$、$\dot{V}_{\mathrm{m}2}$ 和 $\dot{V}_{\mathrm{m}}$ 在 $t=0$ 时刻开始按逆时针方向绕着坐标原点 O 以恒定角速度 ω 旋转。

在任一时刻，旋转矢量 $V_{\mathrm{m}}\angle(\omega t+\varphi)$ 在纵坐标轴 $N'N$ 上的投影等于矢量 $V_{\mathrm{m}1}\angle(\omega t+\varphi_1)$ 和 $V_{\mathrm{m}2}\angle(\omega t+\varphi_2)$ 在同一坐标轴上的投影之和，或者瞬时值 v_1 和 v_2 之和。换句话说，矢量 $V_{\mathrm{m}}\angle(\omega t+\varphi)$ 在纵坐标轴上的投影表示瞬时值之和（v_1+v_2），矢量 $V_{\mathrm{m}}\angle(\omega t+\varphi)$ 表示要求的正弦时间函数 $v=v_1+v_2$。

从阿尔冈图中求出 V_{m} 的长度和角度 φ 后，可将其表示为

$$v=V_m\sin(\omega t + \varphi)$$

下面讨论解析法。对照图 2（略），可写为

$$\dot{V}_{m1} + \dot{V}_{m2} = \dot{V}_m$$

用直角（代数）形式，这些复数可表示为

$$\dot{V}_{m1} = V'_{m1}+jV''_{m1}$$
$$\dot{V}_{m2} = V'_{m2}+jV''_{m2}$$

将其相加，得

$$V'_{m1}+jV''_{m1}+ V'_{m2}+jV''_{m2}= V'_m +jV''_m=\dot{V}_m$$

式中

$$V_m = \sqrt{(V'_m)^2 + (V''_m)^2}$$
$$\tan\varphi =V''_m/ V'_m$$

由于 $\tan\varphi = \tan(\varphi \pm \pi)$，在确定 φ 之前，知道 $\dot{V}_m$ 所在的象限是很重要的。通过函数的实部和虚部的符号能很容易地确定象限。为方便起见，初相角中用角度来表示而不用弧度。

这两种方法可用于任何数目的同频率正弦函数的叠加。

在实际中，人们通常对正弦量的有效值和相位差感兴趣。于是阿尔冈图可通过省略坐标轴而得到简化（它们的位置是不重要的），但矢量之间的相位差并没有改变。此外，不用长度等于正弦量幅值的旋转矢量，而是改变其比例尺，将矢量的长度当作有效值。

在分析中，常做某些处理，把正弦量改成初相角为零的正弦量。同理，不用复数的幅值，而用幅值复数除以 $\sqrt{2}$ 后得到的有效值。为简单起见，有效值复数简单地称为复数电流、复数电压等。

Unit 1.4 Analysis of Small Signal Amplifiers in Mid-Frequency Band

1.4.1 Text

Introduction

In the other lesson, a single small signal equivalent circuit was developed, applicable to both bipolar junction transistors and field effect transistors. This is the transconductance model, and it is redrawn in a more general way in Fig.1 (omitted), where terminal designations 1, 2 and 3 are used, and the numbers corresponding to the various electrodes, as shown in Table 1 (omitted).

As previously explained, the input resistance is virtually an open circuit for the FET, and it is equal to r_{be} for the BJT. The transconductance model is a simplified version of an equivalent circuit known as the hybrid-π model, which takes into account equivalent circuit elements which affect the high frequency performance of the transistor. In the present lesson, frequency effects at both high and low frequencies are ignored, but the simplified analysis is useful over a range of frequencies termed the mid-frequency band, which is the frequency region of interest for general-purpose applications.

General Definitions

Irrespective of the mode of connection of the transistor, amplifier properties can be defined in a general manner, it is essential to observe correctly the polarities of voltages and currents in these definitions, and these are shown in Fig.2 (omitted). Referring to Fig.2 (a), we have

Terminal voltage gain $\qquad A_v \triangleq \dfrac{V_L}{V_i}$ (1)

Voltage gain referred to source $\qquad A_{vs} \triangleq \dfrac{V_L}{V_s}$ (2)

Current gain $\qquad A_i \triangleq \dfrac{I_L}{I_i}$ (3)

Input resistance $\qquad R_i \triangleq \dfrac{V_i}{I_i}$ (4)

Output resistance: this is the internal resistance of the amplifier as seen by the load. It can be found by setting $V_s = 0$, and applying a voltage (in principle) to the output terminals with the load R_L removed. With reference to Fig.2 (b), the output resistance is then

$$R_o \triangleq \dfrac{V_o}{I_o} \qquad (5)$$

It will also be seen from these definitions that

$$A_i = A_v \cdot \frac{R_i}{R_L} \tag{6}$$

and that

$$A_{vs} = A_v \cdot \frac{R_i}{(R_s + R_i)} \tag{7}$$

Amplifier Configurations

The transistor equivalent circuit, Fig.1 (omitted), is seen to have three terminals, while the rather more general block diagram, Fig.2 (omitted), shows the amplifier with four terminals, two inputs and two outputs. This means that one of the transistor terminals must be common to input and output. Any one of the three transistor terminals may be made the common terminal and this gives rise to three distinct amplifier configurations. When the BJT is used, the configurations are known as common Emitter (CE), Common Collector (CC), and Common Base (CB). When the FET is used, the corresponding configurations are the Common Source (CS), the Common Drain (CD), and the Common Gate (CG).

Each circuit configuration has properties best suited to certain applications, and the purpose of this lesson is to examine these properties in detail. At this point, however, it will be useful to have a summary of the main features of each amplifier configuration.

CE and CS Amplifiers

This configuration, common terminal 3, is the most widely used for general purpose amplifier applications. Voltage gains range from moderate to high and the current gain of the CE circuit is also moderate to high. Current gain has little meaning for the CS amplifier and is not specified. This is because the gate input of the FET presents an extremely high resistance and therefore the input current is negligible. The input resistance of the CE amplifier by contrast has values ranging from low to moderate. Both the CS and the CE amplifiers have moderate to high output resistances.

CC and CD Amplifiers

The common terminal 2 configuration is used mainly as a buffer amplifier between a signal source and a low impedance load. As a buffer amplifier, the circuit allows the signal to be transferred to the load while preventing the load from directly affecting the source. The main characteristics of this amplifier configuration are, high to very high input resistance, low to very low output resistance, and a voltage gain which is close to, but always lower than unity【1】. The output voltage "follows" the input voltage, and for this reason the circuits are also known as Emitter Follower, and Source Follower circuits.

CB and CG Amplifiers

The common terminal 1 configuration is used where very good isolation is required between input and output circuits. This is particularly required at high frequencies to prevent oscillations occurring in amplifier circuits. The main circuit properties of this configuration are, moderate to high voltage gain, low input resistance, and high output resistance.

CE and CS Amplifiers

The circuit for the basic common emitter (CE) amplifier is shown in Fig.3 (omitted), and for the common source (CS) amplifier in Fig.4 (omitted). By "common" is meant that the emitter, or source, terminal is common to both input and output signal circuits. In each of the circuits a self-bias resistor, R_E or R_{st}, is shown, but this is effectively short-circuited to signal currents by capacitor C_E or C_s. the DC blocking capacitors C are assumed to be short circuits to signal currents also. The DC supply, V_{CC} or V_{DD}, is also assumed to be a short circuit to signal currents, and in practice this may be achieved by connecting a large capacitor across the supply. In the equivalent circuit, therefore, R_2 appears in parallel with R_L, and R_A appears in parallel with R_B, the latter combination being shown as R_1 in the equivalent circuit of Fig.5 (omitted) 【2】.

All of the important circuit properties can be derived from the equivalent circuit of Fig.6 (omitted).

Input Resistance:

By inspecting Fig.6 (omitted), the input resistance R_i is seen to be

$$R_i = R_1 // r_1 \tag{8}$$

Note that the signal source resistance R_s is not included in the input resistance.

Voltage Gain:

Referring to Fig.6 (omitted) we find that it is convenient to let

$$R_P = r_o // R_2 // R_L \tag{9}$$

For the output loop it is seen that

$$V_L = -g_m V_{13} R_P \tag{10}$$

And by inspection, $V_i = V_{13}$, so that the voltage gain is

$$A_v = \frac{V_L}{V_i} = -g_m R_P \tag{11}$$

Current Gain:

As defined in Section 5.2, the current gain is $A_i = A_v \dfrac{R_i}{R_L}$.

Eq. (11) for voltage gain may, therefore, be substituted in this to give

$$A_i = -g_m R_P \frac{R_i}{R_L} \tag{12}$$

For the FET the input resistance is $R_i = R_1$, and since the signal current in R_1 is negligible, the current gain of the common-source amplifier is not a meaningful quantity. For the BJT, $R_i \approx r_{be}$ and since $g_m r_{be} = \beta$ then $A_i = -\beta \cdot R_P / R_L$.

Output Resistance:

Inspected the circuit in Fig.6 (omitted), the output resistance R_o is

$$R_o = r_o // R_2 \tag{13}$$

Note that R_L is not included in the output resistance.

1.4.2 Specialized English Words

small signal amplifier　小信号放大器
mid-frequency band　中频带

bipolar junction transistor（BJT） 双极型晶体管
field effect transistor（FET） 场效应管
electrode 电极，电焊条
hybrid-π model 混合π形模型
polarity 极性
gain 增益
common base 共基极
common drain 共栅极
self-bias resistor 自偏置电阻
DC supply 直流供电电源
buffer amplifier 缓冲放大器
isolation 隔离，绝缘，隔振
insulator 绝缘体
emitter 发射器，发射极，发射管
emitter follower 射极跟随器
in parallel with 和……并联
parallel circuit 并联电路
parallel resonance 并联谐振
parallel series 混联
common collector 共集电极
common source 共源极
source follower circuit 信号源跟随电路
DC blocking capacitor 直流耦合电容器，隔直流电容器
input resistance 输入电阻
output resistance 输出电阻

1.4.3 Notes

【1】The main characteristics of this amplifier configuration are, high to very high input resistance, low to very low output resistance, and a voltage gain which is close to, but always lower than unity.

该主句的框架是系表结构，主语是 characteristics，表语有三个...input resistance，...output resistance，a voltage gain...，只是前面两个表语分别有一形容词短语 high to very high，low to very low 作为 output resistance 的定语；而最后一个表语的定语则是一个定语从句 which is close to，but always lower than unity，其中 unity=one，所以其意为"（这个电压增益）接近，但总小于1"。

【2】In the equivalent circuit, therefore, R_2 appears in parallel with R_L, and R_A appears in parallel with R_B, the latter combination being shown as R_1 in the equivalent circuit of Fig.

这是一个并列句，其中两个并列的子句分别是 R_2 appears in parallel with R_L，R_A appears in parallel with R_B. 其中 appears 是动词，意为"出现"，翻译时应采用动词的省译法（因为句中含有动作意味的名词——parallel），即将该动词省译，因此译文为"R_2 与 R_L 并联，R_A 与 R_B

并联"。此外，therefore 用两个逗号隔开，插在中间，可视为插入语，但翻译时要译在句首。The latter combination being shown as R_1 in the equivalent circuit of Fig. 是独立分词短语做状语，其中 the latter combination 是逻辑主语，原意为"后一个（即 R_A 与 R_B）并联（物）"，因此将其译为"后一个并联电阻"；being shown as 是独立分词短语的逻辑谓语，译为"表示为……"。

1.4.4 Translation

中频带小信号放大器的分析

引言

在前述课文中，已建立了简单的小信号等效电路，它既适用于双极型晶体管也适用于场效应管。这是一跨导模型，现用更一般的形式重画，如图 1（略）所示。其终端分别用 1、2、3 符号表示，这些数字与表 1（略）所示各电极相对应。如前所述，对场效应管（FET）来说，输入电阻实际上是开路的。而对晶体管（BJT）来说，它等于 r_{be}。跨导模型是通称为混合 π 形电路的等效电路的简化，其考虑了可影响晶体管高频性能的各种等效电路元件。在本文中，高频和低频频效均不予讨论，但这一简单化分析并不影响整个中频范围的应用，而该频带是通常人们感兴趣的频率范围。

定义

不管晶体管的连接方式如何，放大器性能都能用一通用形式来定义。正确观察这些定义中电压和电流的极性是很有必要的，这些极性如图 2（略）所示，从图 2（a）中我们得出：

终端电压增益
$$A_v \triangleq \frac{V_L}{V_i} \tag{1}$$

电源电压增益
$$A_{vs} \triangleq \frac{V_L}{V_s} \tag{2}$$

电流增益
$$A_i \triangleq \frac{I_L}{I_i} \tag{3}$$

输入电阻
$$R_i \triangleq \frac{V_i}{I_i} \tag{4}$$

输出电阻：输出电阻是从负载端看的内电阻。将 V_s 短路，移开负载 R_L 后，在输出端加一电压（原理上），则可求得输出电阻，参看图 2（b），则输出电阻为

$$R_o \triangleq \frac{V_o}{I_o} \tag{5}$$

由定义还可看出

$$A_i = A_v \cdot \frac{R_i}{R_L} \tag{6}$$

和

$$A_{vs} = A_v \cdot \frac{R_i}{(R_s + R_i)} \tag{7}$$

放大器组态

从图 1（略）所示的晶体管等效电路图可看出晶体管有三个端点，而在如图 2（略）所示的更加通用的方框图中，放大器有四个端点、两个输入和两个输出。这就意味着晶体管必须有一个端点是为输入和输出所共用的。晶体管三个终端中的任何一个都可作为共用终端，这就产生了三个截然不同的放大器组态。当使用晶体管时，组态可分为共射极、共集电极和共基极。当使用场效应管时，基本组态是共源极、共漏极和共栅极。

每种组态都有其最适合某种用途的特征，本文的目的就是要详细地分析这些特征。因而，下面对每个放大器组态的主要特征做一概述，这将是很有用的。

CE 和 CS 放大器：这种组态以端点 3 为公共端，是通用放大器中使用最为广泛的线路。电压增益范围由中至高，CE 电路的电流增益也是由中至高。对于 CS 放大器来说，电流增益没有多大意义，因而无此提法。这是因为场效应管（FET）的栅极输入呈现一个极高的电阻，因此，输入电流可以忽略不计。相比之下，CE 放大器的输入电阻值由低至中。CS 和 CE 放大器均有由中至高的输出电阻。

CC 和 CD 放大器：其为共用端点 2 的组态，主要用作信号源与低阻抗负载之间的缓冲放大器。作为缓冲放大器，该线路可允许信号传输给负载，又可防止负载直接影响信号。该放大器组态的主要特征是，高至极高的输入电阻，低至极低的输出电阻，以及具有一个接近但总是小于 1 的电压增益。输出电压"跟随"输入电压，因此，该电路也称为射极跟随器和信号源跟随电路。

CB 和 CG 放大器：其为共用端点 1 的组态，一般在输入和输出线路之间需要很好隔离时使用。在高频状态尤其需要用该线路来防止放大器线路中发生振荡。该线路的主要特征是中至高的电压增益，低的输入电阻和高的输出电阻。

CE 和 CS 放大器

基本的共射放大器（CE）电路和共源放大器（CS）电路如图 3（略）和图 4（略）所示。"公共"的意思是发射极或源极的端头为输入和输出信号电路所共用。如图所示，每个电路中均有自偏置电阻，R_E 和 R_{st}，但其信号电流被电容 C_E 或 C_s 有效地短路了。直流隔离电容 C 被认为将信号电流短路了。直流供电电源 V_{CC} 或 V_{DD}，也被认为对信号电流是短路的，在实际中，只要将一大电容与电源并联便可实现。因此，在等效电路中，R_2 与 R_L 并联，R_A 与 R_B 并联电阻在等效电路图（如图 5（略）所示）中表示为 R_1。

由等效电路图（如图 6（略）所示）可以得出所有重要的电路特征。

输入电阻：

由图 6（略）可知，输出电阻 R_i 为

$$R_i = R_1 // r_1 \tag{8}$$

注意信号源电阻 R_s 未包括在输入电阻中。

电压增益：

由图 6（略）可以看出，令

$$R_P = r_o // R_2 // R_L \tag{9}$$

是方便的。

对于输出回路，可以得出

$$V_L = -g_m V_{13} R_P \tag{10}$$

经观察，$V_i = V_{13}$，所以电压增益为

$$A_i = \frac{V_L}{V_i} = -g_m R_P \tag{11}$$

电流增益：

如前所定义，电流增益为
$$A_i = A_v \frac{R_i}{R_L}$$

因此，将电压增益公式（11）代入上式中，则得到

$$A_i = -g_m R_P \frac{R_i}{R_L} \tag{12}$$

对场效应管（FET）来说，输入电阻是 $R_i = R_1$，由于 R_1 中的信号电流可以忽略不计，这样共源放大器的电流增益定量计算没有意义了。对于晶体管（BJT）来说，$R_i \approx r_{be}$，由于 $g_m r_{be} = \beta$，因此，$A_i = -\beta \cdot R_P / R_L$。

输出电阻：

由图6（略）可以看出，输出电阻 R_o 为

$$R_o = r_o // R_2 \tag{13}$$

注意 R_L 未包括在输出电阻内。

Unit 1.5 Logic Circuits

1.5.1 Text

Introduction

Logic circuits are used in building computers of all types, ranging from the very largest ones down to the hand-held calculator, as well as a wide variety of other electronic systems 【1】. While all of these systems can be built up using only a few basic, simple building block circuits called gates, many systems and subsystems are now being produced in single integrated circuits of the large-scale (LSI) and very large-scale (VLSI) types, each of which may contain many thousands of the simple gate circuits 【2】. Digital circuits and techniques are indeed taking over many areas that have traditionally used linear techniques, especially in areas where microcomputers may effectively be used.

This lesson presents the basic logic gate circuits which serve as the building blocks for more complex systems, and shows how electronic devices such as diodes and transistors can be used to realize these gate circuits. Several of the more popular logic circuit families are discussed and the more important characteristics of each are presented. This lesson is not intended to be a complete course in logic circuits, and assumes that the reader has already become familiar with logic and its use from another text, such as *Digital Computer Circuits and Concepts* by Deem, Muchow, and Zeppa (Reston Publishing Co.1980). Rather, the material presented here should provide a supplement to material presented in such a text.

Digital Logic

Computational arithmetic is based on the decimal or Arabic number system, which uses ten states for each variable, designated by the digits 0 through 9 【3】. However, in modern computers the arithmetic processes are all performed using binary or two-state arithmetic. This is so because the devices used to perform the arithmetic or logic functions are neatly all binary in nature. It is possible to create certain types of multiple-state devices, but the binary ones are the easiest and least expensive to build. Thus the numbers on which a computer is to operate are first converted to binary combinations, the arithmetic or logic operations are performed by the machine in binary form, and the results are then converted back to decimal numbers at the output.

In the binary system, the two states are represented by the two logic symbols 0 or 1, corresponding to the symbols 0 through 9 used in the decimal system. These may be thought of in the same sense as combinations of false or true, or of off or on. In computer systems the logic levels 0 or 1 most often are represented by levels of voltage, where a low voltage is a logic 0 and a high voltage is a logic 1.

A logic variable is simply the logic condition at a certain point in a logic circuit. It may be for instance the logic condition at one of the input leads, or it may be the logic condition at the output lead, or it may be a logic condition at some internal point. The logic variable is given name or symbol to distinguish it from all other variables. Each variable in the system may only take on one of two conditions, a logic 0 or a logic 1. For example, a simple gating circuit may have two input variables A and B, and one output variable X. Each of A and B may take on the values 0 or 1 independently of each other, but the output X may only take on a value if certain combinations of A and B exist. Thus, X is a logic function of A and B, which is determined by the connections within the gating circuit.

The entire system of Boolean algebra can be built up using combinations of only three basic functions, the logic AND, the logic OR, and the logic NOT (or negation). Any Boolean equation can be realized by some combination of these three basic functions. These three basic functions will be discussed in detail below.

A truth table is a list of all of the possible input variable state combinations of a circuit listed in binary-sequential order with the corresponding output state for each combination listed in an adjacent column. Table1 (omitted) shows the binary numbers corresponding to the decimal numbers from zero to fifteen.

The truth table is used as the beginning point in designing or analyzing a logic circuit. The sequential listing makes it easy to recognize if any input combinations were missed 【4】. It is made up either from the problem specifications or by sequential testing of an assembled circuit. A logic equation can be formulated from the truth table and a logic circuit can be developed from the equation. Truth tables are used in defining the basic AND, OR and NOT functions below.

Primary Logic Functions

(1) Logic AND Function

The logic AND function is defined as follows: "An AND function will have a logic 1 output if and only if all of its input variables are in the logic 1 state." In True/False terminology, this says that the output will be true only if all of the inputs are true, and the output will be false if any of the inputs are false.

The logic symbol used to represent the AND function is shown in Fig.1 (omitted), for the two-input case. The two input variables are A and B, and the output X is the logic AND combination of these inputs. This is expressed algebraically by the equation

$$X = A \cdot B = AB \tag{1}$$

where the dot indicates the logic AND operation and is analogous to multiplication in simple algebra. As in simple algebra, it is common practice to omit the dot, with the AND function implied by the combination AB.

The truth table for the logic AND function is show in Fig.2 (omitted). As noted in the definition, the only combination of AB which produces a 1 output is that for which $AB=11$. All the others produce a 0 output.

The logic AND function can be expanded to any number of inputs, simply by applying the basic definition. Thus a 5-input logic AND gate will produce an output only if the input combination $ABCDE$=11111 exists, and providing the physical circuit will still function properly with that many inputs attached 【5】.

(2) Logic OR Function

The logic OR function, more precisely called the inclusive OR function, is defined by the following statement. "An OR function will have a logic 1 output if any or all of its inputs have a logic 1 state, and a logic 0 output if and only if all of its inputs are logic 0."

The logic symbol for the OR function i shown in Fig.3(omitted), for the two-input case, where A and B are the two inputs and X is the output. The algebraic equation for the two-input OR function is given by

$$X=A+B \qquad (2)$$

where the "+" sign indicates the logic OR operation and is analogous to addition in ordinary algebra.

The truth table for the logic OR function is shown in Fig.4 (omitted), and as noted in the definition, a 1 output occurs for any combination of AB where one or more of the inputs is 1, and 0 occurs when all inputs are 0 【6】.

The OR function can also be expanded to any number of inputs, again by applying the basic definition. Thus a 5-input OR gate will produce a logic 0 output only when $ABCDE$=00000, and will produce a logic 1 output for all other combinations.

(3) Logic NOT Function

The logic NOT function is defined by the following statement. "The output of a NOT function will be a logic 1 when its input is a logic 0 and the output will be a logic 0 when its input is a logic 1. The logic NOT function can have only one input."

The logic symbol and logic equation for the NOT function are shown in Fig.5 (omitted). The symbol is comprised of an amplifier triangle symbol combined with a small circle on its output indicating the negation operation. The small circle may also be used in the input or output leads to AND and OR gates to indicate negation of the associated variables. In the logic equation, a bar line above the variable symbol indicates the negation. The bar line may be extended to cover an entire expression as shown in the third case of Fig.6 (omitted). In this case the output X is the negation of Y, which is the output of the AND function AB, or

$$X=\overline{Y}=\overline{AB} \quad \text{where } Y=AB \qquad (3)$$

(4) Logic NOR function

The NOR or negated-OR function is sometimes called the "universal gate", since as will be seen below any Boolean equation can be realized using only NOR gates. The NOR function is defined as a circuit that produces output only if none of the inputs are 1, that is just the inverse of the OR function. The truth table and logic symbol for the NOR gate are shown in Fig.7 (omitted), and the equation for the output is

$$X=\overline{A+B} \qquad (4)$$

The NOR gate may be used to perform the logic OR function by passing its output through an inverter to reinvent it. This is easily seen because the NOR function is defined as an OR function followed by an inversion (NOT), and two inversions cancel each other in the manner of double negation.

The NOR gate may also be used to perform the logic AND function by inverting the inputs before they are applied to the NOR gate. This result is not so easily seen, but may be proved by use of a technique called De Morgan's Theorem, or by filling out a truth table with columns for A, B, $\overline{A}$, $\overline{B}$, and the output which is the NOT of $\overline{A}$, $\overline{B}$.

NOR gates may also be used to perform the NOT function. For example, two inputs of a NOR gate are tied together to a common input A, so that the output is the inversion of A OR A, which is just the inversion of A and the second input of the NOR gate is tied to a point that is always at a logic 0 level, such as the negative power supply voltage in TTL logic[7]. Writing the truth table for A NOR 0 will quickly show that the output is indeed $\overline{A}$.

However, if the NOR gate is connected with its second input to a logic 1 level, the output will be stuck at a logic 0 level, and the result is not a logic function at all.

Now, since all three of the basic logic functions, AND, OR and NOT can be realized using only NOR gates, the NOR gate is itself a basic function from which all logic functions can be realized. It is therefore called a universal gate.

1.5.2 Specialized English Words

logic circuit　逻辑电路
hand-held calculator　便携式计数器
building-block circuit　积木式结构电路
subsystem　子系统，辅助系统
intergrated circuit of the large-scale　大规模集成电路
very large-scale Intergration（VLSI）　超大规模集成电路
computational arithmetic　算术运算
decimal system　十进制
Arabic number　阿拉伯计数制
chance variable　随机变量
dependent variable　因变量，函数
independent variable　自变量
binary　二进制
simple algebra　初等代数
logic OR function　逻辑"或"函数
De Morgan's Theorem　德·摩根定理（德·摩根是19世纪英国数学家）
binary combination　二进制数组
decimal number　十进制数
logic symbol　逻辑符号
logic variable　逻辑变量
logic condition　逻辑状态
output lead　输出端

Boolean algebra　布尔代数
binary-sequential order　二进制次序
truth table　真值表
assembled circuit　集成电路
logic AND function　逻辑"与"函数
terminology　术语，专门名词
multiple-state　多态
logic equation　逻辑方程
universal gate　全能门

1.5.3　Notes

【1】Logic circuits are used in building computers of all types, ranging from the very largest ones down to the hand-held calculator, as well as a wide variety of other electronic systems.

　　该句的句干为 Logic circuits are used in building computers of all types...as well as a wide variety of other electronic systems. ranging from the very largest ones down to the hand-held calculator 是由现在分词 ranging 引导的分词短语，修饰 computers of all types，说明其"范围大至巨型计算机，小至便携式计算器"。

【2】While all of these systems can be built up using only a few basic, simple building block circuits called gates, many systems and subsystems are now being produced in single integrated circuits of the large-scale(LSI)and very large-scale(VLSI)types, each of which may contain ma-ny thousands of the simple gate circuits.

　　该句的句干为 While all of these systems can be built up using...building block circuits...，many systems and subsystems are now being produced in single integrated circuits...，只是在状语从句中，分词 using 的宾语 building block circuits 有三个定语，两个前置定语 a few basic, simple 用逗号","分开，一个过去分词短语 called gates 做后置定语；而在主句中，介词 in 的宾语 integrated circuits 也有两个定语，一个是由 of 引导的介词短语 of the large-scale（LSI）and very large-scale（VLSI）types 做后置定语；另一个是由 each of which 引导的非限定性定语从句 each of which may contain many thousands of the simple gate circuits。

【3】Computational arithmetic is based on the decimal or Arabic number system, which uses ten states for each variable, designated by the digits 0 through 9.

　　句中由 which 引导的非限定性定语从句中动词 uses 的用法是 use sth. for sth.，for each variable 是状语，而 designated by the digits 0 through 9 是过去分词短语做 ten states 的后置定语，但因其比较长，放在句末，并用逗号将其分隔，这就是语法中常说的做定语的分词短语与其被修饰词分隔的现象。弄清了以上关系，本句就不难翻译了。

【4】The sequential listing makes it easy to recognize if any input combinations were missed.

　　本句的谓语动词的用法仍然是 sth.（subject）make sth.（object）adj. 的形式，只是此处宾语是动词不定式 to recognize if any input combinations were missed，因为较长，为使句子不"头重脚轻"，因此用 it 作为其先行词，而将整个不定式放在句末。还应看出，此处从 if 一直到句末为一从句，做不定式 to recognize 的宾语。

【5】Thus a 5-input logic AND gate will produce an output only if the input combination $ABCDE$=11111 exists, and providing the physical circuit will still function properly with that many inputs attached.

首先应看出，该句的主句是 a 5-input logic AND gate will produce an output（Thus 是连接上句的语气连词）。只是"一个五输入的逻辑'与'门输出'1'是有条件的"，就是分别由 if 和 providing（that）这两个从属连词引导的条件状语从句。其中前一从句的主语是 input combination $ABCDE$=11111 这样一个短语表达式（其中 input combination 与 $ABCDE$ 是同位语），exists 是谓语；后一从句的谓语是动词 function，attached 意为"连接的"，做 inputs 的后置定语。

【6】The truth table for the logic OR function is shown in Fig.3（omitted），and as noted in the definition, a 1 output occurs for any combination of AB where one or more of the inputs is 1, and 0 occurs when all inputs are 0.

整句总框架是一并列句，The truth table for the logic OR function is shown in Fig.3 是前一子句，and 以后直至句末为第二个子句；后一子句本身又是一个并列句，其句干为 a 1 output occurs...and 0 occurs...，而 as noted in the definition 是分词短语做该并列句的状语。前一子句中，where one or more of the inputs is 1 是修饰 any combination of AB 的定语从句；后一子句中 when all inputs are 0 是其时间状语从句。

【7】For example, two inputs of a NOR gate are tied together to a common input A, so that the output is the inversion of A OR A, which is just the inversion of A and the second input of the NOR gate is tied to a point that is always at a logic 0 level, such as the negative power supply voltage in TTL logic.

本句由插入语 For example 引出两个例子，这两个例子分别由两个并列的子句叙述。前一子句是 two inputs of a NOR gate are tied..., which is just the inversion of A，后一子句从 the second input of...一直到句末。前一子句本身是一主从复合句，two inputs of a NOR gate are tied together to a common input A，意为"将'或非'门的两个输入端一起接到同一输入端 A"，其结果便是由 so that 引导的结果状语从句所叙述的内容：so that the output is the inversion of A OR A, which is just the inversion of A 意为"这样，其输出便为 $A+A$ 的'非'，这正是 A 的'非'"。后一子句的主句是 the second input of the NOR gate is tied to a point；that is always at a logic 0 level, such as the negative power supply voltage in TTL logic. 是定语从句，修饰 a point，其中 such as the negative power supply voltage in TTL logic 是介词短语，进一步说明 a point 意为"TTL 逻辑电路中的电源负端"。

1.5.4　Translation

逻 辑 电 路

引言

　　逻辑电路用于构造各种类型的计算机（大到巨型计算机，小到便携式计数器），以及大量的其他电子系统。尽管所有这些系统可只用几种基本的、简单的称之为"门"的积木式电路来构造，但现在许多系统和辅助系统正在用单个的大规模以及超大规模集成电路来实现。其中每一块可能包括成千上万个简单的门电路。数字电路及技术确实在占领许多传统使用线性技术的领域，尤其是那些有效使用微型计算机的领域。

本文讲述构成较复杂系统的基本部件的一些逻辑门电路，阐明了诸如二极管和晶体管这些电子器件是如何被用来实现这些门电路的，讨论几种比较常见的逻辑电路，一一介绍它们的一些重要特性。本文不打算完整地介绍逻辑电路，并认为读者已从其他课本，例如迪姆、马歇尔和齐柏著述的《数字电子计算机电路与概念》（赖斯顿出版社，1980年版）的学习，对逻辑电路以及它的使用已比较熟悉。更确切地说，这里介绍的内容是对那些内容的补充。

数字逻辑

算术运算建立在十进制数即阿拉伯数字系统上，其用0~9这10个数字所定义的10种状态表示每个变量。然而在现代计算机的算术运算过程中，全部使用二进制运算。这是因为用来完成算术或逻辑功能的设备，本身几乎全部是二进制的。构造某种多状态设备是可能的，但二进制设备建造起来是最简易也是最经济的。这样计算机要进行操作的数字必须先转换成二进制数，算术或逻辑操作以二进制进行，最后运算结束再转换回十进制数结果输出。

与十进制中的0~9相对应，在二进制系统中，用0或1这两个逻辑符号来表示两种状态。在一定意义上，0和1可以认为是错或对，或者是开或关的一对组合。在计算机系统中，逻辑0或1最常见的是用电平来表示，低电平为逻辑0，高电平为逻辑1。

逻辑变量只是逻辑电路中某一点的逻辑状态。例如，它可能是一个输入端的逻辑状态，也可能是输出端的逻辑状态，也可能是一些内部节点的逻辑状态。逻辑变量用一个专门的名字或符号来表示，以与所有其他变量有所区别。系统中的每一个变量只能以两种逻辑状态"0"或"1"中的一种形式出现。例如，一个简单的门电路可能有两个输入变量A和B，以及一个输出变量X。A和B都可相互独立地取0或1，但输出变量X却只能在A和B有某种组合关系时才呈现某值。这样X为A和B的逻辑函数，它由门电路内部的关系所决定。

布尔代数的整个系统都可由三个基本函数即逻辑"与"、"或"和"非"组合构成。任何一个布尔等式可由这三个基本函数中的一部分组合而成，以下将详细讨论这三个基本函数。真值表是这样的一个表格，电路中可能输入的所有不同的状态组合按二进制顺序连续排列，输出状态与输入端的每种组合一一对应，表1（略）表示了与十进制数0~15一一对应的二进制数。

在设计或分析一个逻辑电路时，首先要列写真值表。这种连续的表格很容易辨认出是否遗漏了任何一种组合，它可由问题的具体情况或对一集成电路进行连续测试来建立。由真值表可建立逻辑等式，由此等式就可构造一逻辑电路。在下面定义基本逻辑"与"、"或"和"非"的函数中就用到了真值表。

基本逻辑函数

（1）逻辑"与"函数

逻辑"与"函数定义如下："与"函数有且只有在它的输入变量全为逻辑"1"状态时才输出"1"。在"真"与"假"的术语中，这可表述为如果输入全为"真"则输出为真，如果有任一输入为"假"则输出为"假"。

对于两输入的情况，表达"与"的逻辑符号如图1（略）所示。假定两输入变量是A和B，则输出X是这两个输入的逻辑"与"组合，可用数学等式来表达

$$X = A \cdot B = AB \tag{1}$$

式中，"·"代表逻辑"与"运算，它与初等代数中的"乘"类似。就像在初等代数中常省略那样，"与"函数通常用 AB 表示。

逻辑"与"的真值表如图 2（略）所示。正如定义的那样，只有在 $AB=11$，输出才为 1。所有其他情况都输出为 0。

只要运用基本定义，逻辑"与"函数便可扩展到任意一个输入。因此，一个五输入的逻辑"与"门只有在 $ABCDE=11111$ 且具体电路与那些连接的输入量有恰当的函数关系时才等于 1。

(2) 逻辑"或"函数

逻辑"或"函数，准确地称之为"广义或"函数，定义如下："或"函数在输入变量中任一个为 1 时输出为 1，只有在输入全为 0 时才输出 0。

对于两输入情况，"或"函数的逻辑符号如图 3（略）所示，图中 A 和 B 代表两个输入变量，X 代表输出。两输入的"或"函数的代数方程为

$$X = A + B \tag{2}$$

式中，"+"号代表逻辑或运算，与普通代数中加法相类似。

逻辑"或"的真值表如图 4（略）所示。如定义所指出的那样，AB 中有一个或两个为 1 时，输出为 1，AB 全为 0 时，输出才为 0。

"或"函数也可通过运用基本定义扩展为任意一个输入端。因此，5 个输入端"或"门只有在 $ABCDE=00000$ 时才输出 0，对其他任意输入组合都输出 1。

(3) 逻辑"非"函数

逻辑"非"函数定义如下："非"函数当输入为逻辑 0 时输出 1，输入 1 时输出 0，逻辑"非"的函数只有一个输入变量。

逻辑"非"符号和表达式如图 5（略）所示。符号由一个放大器的三角形符号和在一个输出端代表"非"运算的小圆圈组成。小圆圈也可用在"与"门和"或"门的输入或输出端和"与"和"或"门相连的连线上，以表示几个相互并联变量的"非"。在逻辑表达式中，变量符号上的横线代表"非"，横线也可用在整个表达式上，如图 6（略）所示的三种情况。在这种情形中，输出 X 是 Y 的非，Y 又是 AB 与函数的输出，即

$$X = \overline{Y} = \overline{AB} \tag{3}$$

式中，$Y = AB$。

(4) 逻辑"或非"函数

从下面可以看出，由于任何布尔表达式都可由"或非"门实现，因此"或非"函数有时又称为"全能门"。"或非"门电路定义为：只有当输入变量没有 1 时才输出 1，也就是"或"函数的非。"或非"的真值表和逻辑符号如图 7（略）所示，其输出表达式为

$$X = \overline{A + B} \tag{4}$$

在"或非"门的输出端加上一"非"（将输出变量再反相）便可实现逻辑或功能。"或非"函数定义为由"或"函数加上"非"函数，这一点很容易看出，两个"非"在双重否定时彼此抵消。

"或非"也可用来在两变量输入它之前，加上两个"非"门来构成"与"门。这一结论不是那么容易看出，但却可以用德·摩根定理或者列出真值表来证明，表内列出 A，B，$\overline{A}$，$\overline{B}$ 以及 $\overline{A}$，$\overline{B}$ 的或非输出。

"或非"也可用来实现"非"的功能，例如，将"或非"门的两个输入端一起接到 A，这

样其输出便为 $A+A$ 的"非",这正是 A 的非;或者将"或非"第二输入端连接到逻辑电平恒为 0 的那一点,例如连接到 TTL 逻辑门中的负电源电压(也就是地)那一点。写出 A 或 0 的真值表便可看出输出的确为 $\overline{A}$。

然而,假如"或非"门的第二输入端连接到逻辑"1",输出端将恒为"0",其结果根本不是一个逻辑函数。

由于三种基本的逻辑函数"与"、"或"和"非"都可由"或非"门构成,因此"或非"门本身就是一个可实现各种逻辑功能的基本函数,因此它被称为"全能门"。

Part 2　Electric Machine & Its Control

Unit 2.1　Transformer Principle

Unit 2.2　D.C. Machine Construction and Principle

Unit 2.3　Three-Phase Induction Machine Principle

Unit 2.4　Operation Analysis of Three-Phase Synchronous Generator

Unit 2.5　Motor Control Based on Microcomputer

Unit 2.1　Transformer Principle

2.1.1　Text

It has been shown that a primary input voltage V_1 can be transformed to any desired open-circuit secondary voltage E_2 by a suitable choice of turn ratio. E_2 is available for circulating a load current whose magnitude and power factor will be determined by the secondary circuit impedance. For the moment, a lagging power factor will be considered. The secondary current and the resulting ampere-turns I_2N_2 will change the flux, tending to demagnetize the core, reduce Φ_m and with it E_1. Because the primary leakage impedance drop is so low, a small alteration to E_1 will cause an appreciable increase of primary current from I_0 to a new value of I_1 equal to $(V_1+E_1)/(R_1+jX_1)$. The extra primary current and ampere-turns nearly cancel the whole of the secondary ampere-turns. This being so, the mutual flux suffers only a slight modification and requires practically the same net ampere-turns I_0N_1 as on no load. The total primary ampere-turns are increased by an amount I_2N_2 necessary to neutralize the same amount of secondary ampere-turns [1]. In the vector equation, $\dot{I}_1N_1 + \dot{I}_2N_2 = \dot{I}_0N_1$; alternatively, $\dot{I}_1N_1 = \dot{I}_0N_1 - \dot{I}_2N_2$. At full load, the current I_0 is only about 5% of the full-load current and so I_1 is nearly equal to I_2N_2/N_1. Bearing in mind that $E_1=E_2N_1/N_2$, the input kVA which is approximately E_1I_1 is also approximately equal to the output kVA, E_2I_2.

Neglecting the resistance for the moment, the physical picture will now be considered with reference to Fig.1(omitted). The primary current has increased, and with it the primary leakage flux to which it is proportional [2]. The total flux linking the primary, $\Phi_p=\Phi_m+\Phi_1=\Phi_{11}$, is shown unchanged because the total back e.m.f., $E_1-N_1d\Phi_1/dt$ is still equal and opposite to V_1. However, there has been a redistribution of flux and the mutual component has fallen due to the increase of Φ_1 with I_1. Although the change is small, the secondary demand could not be met without a mutual flux and e.m.f. alteration to permit primary current to change. The net flux Φ_s linking the secondary winding has been further reduced by the establishment of secondary leakage flux due to I_2, and this oppose Φ_m [3]. Although Φ_m and Φ_2 are indicated separately, they combine to one resultant in the core which will be downwards at the instant shown. Thus the secondary terminal voltage is reduced to $V_2=-N_2d\Phi_s/dt$ which can be considered in two components, i.e. $V_2=-N_2d\Phi_m/dt-N_2d\Phi_2/dt$ or vectorially $=E_2-jX_2I_2$. As for the primary, Φ_2 is responsible for a substantially constant secondary leakage inductance $N_2\Phi_2/i_2=N_2^2\Lambda_2$. It will be noticed that the primary leakage flux is responsible for part of the change in the secondary terminal voltage due to its effects on the mutual flux. The two leakage fluxes are closely related; Φ_1, for example, by its demagnetizing action on Φ_m has caused the changes on the primary side which led to the establishment of primary leakage flux.

As previously explained, it is usual to treat the leakage e.m.f.s in terms of the voltage drops in antiphase. Fig.2 (omitted) shows all the flux components and e.m.f.s and should be compared with the figure of the no load transformer, which shows the voltage drop components and includes the small resistance drops.

So far, a lagging power factor has been assumed. If a low enough leading power factor is considered, the total secondary flux and the mutual flux are increased causing the secondary terminal voltage to rise with load. Φ_p is unchanged in magnitude from the no load condition since, neglecting resistance, it still has to provide a total back e.m.f. equal to V_1. It is virtually the same as Φ_{11}, though now produced by the combined effect of primary and secondary ampere-turns【4】. The mutual flux must still change with load to give a change of E_1 and permit more primary current to flow 【5】. E_1 has increased this time but due to the vector combination with V_1 there is still an increase of primary current.

Two more points should be made about the figures. Firstly, a unity turns ratio has been assumed for convenience so that $E_1=E_2'$. Secondly, the physical picture is drawn for a different instant of time from the vector diagrams which show $\Phi_m=0$, if the horizontal axis is taken as usual, to be the zero time reference. There are instants in the cycle when primary leakage flux is zero, when the secondary leakage flux is zero, and when primary and secondary leakage fluxes are in the same sense【6】.

Equivalent Circuit

The equivalent circuit already derived for the transformer with the secondary terminals open, can easily be extended to cover the loaded secondary by the addition of the secondary resistance and leakage reactance.

Practically all transformers have a turn ratio different from unity although such an arrangement is sometimes employed for the purposes of electrically isolating one circuit from another operating at the same voltage. To explain the case where $N_1 \neq N_2$ the reaction of the secondary will be viewed from the primary winding. The reaction is experienced only in terms of the magnetizing force due to the secondary ampere-turns. There is no way of detecting from the primary side whether I_2 is large and N_2 small or vice versa, it is the product of current and turns which causes the reaction【7】. Consequently, a secondary winding can be replaced by any number of different equivalent windings and load circuits which will give rise to an identical reaction on the primary. It is clearly convenient to change the secondary winding to an equivalent winding having the same number of turns N_1 as the primary. The circuit of Fig.3 (omitted) would then be applicable although changes from the actual secondary winding parameters would be expected.

E_2 to E_2', I_2 to I_2', and Z_2 to Z_2' because N_2 changes to $N_2'=N_1$.

With N_2 changes to N_1, since the e.m.f.s are proportional to turns, $E_2'=E_2(N_1/N_2)$ which is the same as E_1.

For current, since the reaction ampere turns must be unchanged $I_2'N_2'=I_2'N_1$ must be equal to I_2N_2. i.e. $I_2'=(N_2/N_1)I_2$.

For impedance, since any secondary voltage V becomes $(N_1/N_2)V$, and secondary current I becomes $(N_2/N_1)I$, then any secondary impedance, including load impedance, must become $V'/I'=(N_1/N_2)^2 V/I$. Consequently, $R_2'=(N_1/N_2)^2 R_2$ and $X_2'=(N_1/N_2)^2 X_2$.

If the primary turns are taken as reference turns, the process is called referring to the primary side.

There are a few checks which can be made to see if the procedure outlined is valid.

For example, the copper loss in the referred secondary winding must be the same as in the original secondary otherwise the primary would have to supply a different loss power.

$I_2'^2 R_2'$ must be equal to $I_2^2 R_2$.

$\left(I_2 \cdot \dfrac{N_2}{N_1}\right)^2 \left(R_2 \cdot \dfrac{N_1^2}{N_2^2}\right)$ does in fact reduce to $I_2^2 R_2$.

Similarly the stored magnetic energy in the leakage field $\left(\dfrac{1}{2} \cdot LI^2\right)$ which is proportional to $I_2^2 X_2$ will be found to check as $I_2'^2 X_2'$ Referred Secondary

$$kVA = E_2' I_2' = E_2(N_1/N_2) \cdot I_2(N_2/N_1) = E_2 I_2$$

The argument is sound, though at first it may have seemed suspect. In fact, if the actual secondary winding was removed physically from the core and replaced by the equivalent winding and load circuit designed to give the parameters N_1, R_2', X_2' and I_2'; measurements from the primary terminals would be unable to detect any difference in secondary ampere-turns, kVA demand or copper loss, under normal power frequency operation.

There is no point in choosing any basis other than equal turns on primary and referred secondary, but it is sometimes convenient to refer the primary to the secondary winding. In this case, if all the subscript 1's are interchanged for the subscript 2's, the necessary referring constants are easily found; e.g. $R_1' = R_1(N_2/N_1)^2$. It is worth noting that for a practical transformer, $R_1' \approx R_2$, $X_1' \approx X_2$; similarly $R_2' \approx R_1$ and $X_2' \approx X_1$.

The equivalent circuit for the general case where $N_1 \neq N_2$ is the same as Fig.4 (omitted) except that r_m has been added to allow for iron loss and an ideal lossless transformation has been included before the secondary terminals to return V_2' to V_2. All calculations of internal voltage and power losses are made before this ideal transformation is applied. The behavior of a transformer as detected at both sets of terminals is the same as the behavior detected at the corresponding terminals of this circuit when the appropriate parameters are inserted. The slightly different representation showing the coils N_1 and N_2 side by side with a core in between is only used for convenience. On the transformer itself, the coils are, of course, wound round the same core.

Very little error is introduced if the magnetizing branch is transferred to the primary terminals, but a few anomalies will arise 【8】. For example, the current shown flowing through the primary impedance is no longer the whole of the primary current. The error is quite small since I_0 is usually such a small fraction of I_1. Slightly different answers may be obtained to a particular problem depending on whether or not allowance is made for this error. With this simplified circuit, the primary and referred secondary impedances can be added to give:

$$R_{e1} = R_1 + R_2(N_1/N_2)^2 \text{ and } X_{e1} = X_1 + X_2(N_1/N_2)^2$$

It should be pointed out that the equivalent circuit as derived here is only valid for normal operation at power frequencies; capacitance effects must be taken into account whenever the rate of change of voltage would give rise to appreciable capacitance currents, $I_C=CdV/dt$【9】. These are important at high voltages and at frequencies much beyond 100 cycles/sec. A further point is that Fig.5（omitted）is not the only possible equivalent circuit even for power frequencies. An alternative, treating the transformer as a three or four-terminal network gives rise to a representation which is just as accurate and has some advantages for the circuit engineer who treats all devices as circuit elements with certain transfer properties【10】. The circuit on this basis would have a turns ratio having a phase shift as well as a magnitude change, and the impedances would not be the same as those of the windings. The circuit would not explain the phenomena within the device like the effects of saturation, so for an understanding of internal behavior, Fig.5（omitted）is preferable.

There are two ways of looking at the equivalent circuit:

（a）viewed from the primary as a sink but the referred load impedance connected across V_2'.

（b）viewed from the secondary as a source of constant voltage V_1 with internal drops due to R_{e1} and X_{e1}.

The magnetizing branch is sometimes omitted in this representation and so the circuit reduces to a generator producing a constant voltage E_1（actually equal to V_1）and having an internal impedance $R+jX$（actually equal to R_e+jX_e）.

In either case, the parameters could be referred to the secondary winding and this may save calculation time.

The resistances and reactances can be obtained from two simple light load tests.

The importance of understanding this preliminary transformer work cannot be overemphasized【11】. It has bearings on the whole of machine behavior and Fig.6（omitted）, for example, will form the basis for the induction machine equivalent circuit.

2.1.2 Specialized English Words

a lagging power factor　滞后的功率因数
in antiphase　反相
stored magnetic energy　磁场储能
capacitance effect　电容效应
sink　（倒）U形（电路）
demagnetize　去磁作用
equivalent circuit　等效电路，等值电路
rate of change of voltage　电压变化率
phase shift　相位移

2.1.3 Notes

【1】The total primary ampere-turns are increased by an amount I_2N_2 necessary to neutralize the same amount of secondary ampere-turns.

句中有"形容词＋不定式"短语做 I_2N_2 的后置定语，也可看成一定语从句 which is

necessary to neutralize the same amount of secondary ampere-turns 的变形。

【2】The primary current has increased, and with it the primary leakage flux to which it is proportional.

此句已经有省略,完整的全句应为:The primary current has increased, and with it the primary leakage flux to which it is proportional has also increased. 因此全句为由 and 连接的并列句, with it 为介词短语,做后一个子句的状语;这前一个 it 是指 primary current;(to) which it is proportional 是定语从句,做前一个 it 的定语;从句中的第二个 it 还是指 The primary current。

【3】The net flux Φ_s linking the secondary winding has been further reduced by the establishment of secondary leakage flux due to I_2, and this oppose Φ_m.

句中的 this 指代的是 the establishment of secondary leakage flux,意为"二次侧的漏磁通的(建立)方向是与 Φ_m 相反的"。

【4】It is virtually the same as Φ_{11}, though now produced by the combined effect of primary and secondary ampere-turns.

句中 it 指代上句中的 Φ_p。though now produced by...可扩展成整句 though now it is produced by...这里的 it 作用同上。

【5】The mutual flux must still change with load to give a change of E_1 and permit more primary current to flow.

句子中由介词 to 引导的动词不定式定语做目的状语,其中实际上有两个并列的不定式,即(to)give a change of E_1 和(to)permit more primary current to flow,两者用连词 and 连接。

【6】There are instants in the cycle when primary leakage flux is zero, when the secondary leakage flux is zero, and when primary and secondary leakage fluxes are in the same sense.

该句中有三个用 when 引导的并列的定语从句,均修饰 instants。

【7】There is no way of detecting from the primary side whether I_2 is large and N_2 small or vice versa, it is the product of current and turns which causes the reaction.

句中, There is no way...是一种常用表达法,意为"无法……"; whether...or ..., 意为"是……还是……", 在句中做动名词 detecting 的宾语;介词短语 from the primary side 做状语;句子 it is the product of current and turns which causes the reaction 是一个典型的强调句型,所强调的部分是 the product of current and turns,该部分在从句中是主语,即强调主语部分,其正常句型为 the product of current and turns causes the reaction,译为"正是电流与匝数的乘积在产生作用";vice versa 译为"反之亦然"。

【8】Very little error is introduced if the magnetizing branch is transferred to the primary terminals, but a few anomalies will arise.

在翻译该句时,要注意否定与肯定、消极与积极的区别,即 little 与 a little、few 与 a few 的区别。little 意为"没什么,很小",a little 意为"有一些",即第一句中表示"如果将激磁支路移至一次侧端口,没有什么误差会引入",而第二句却为一个转折"会有一些不合常理的现象出现",两句正好用 but 将语气转折。

【9】It should be pointed out that the equivalent circuit as derived here is only valid for normal operation at power frequencies; capacitance effects must be taken into account whenever the rate of change of voltage would give rise to appreciable capacitance currents, $I_C = CdV/dt$.

这个句子是典型的主语从句，主句主干是 It should be pointed out，It 做形式主语，that 引导的后面部分内容为真正主语，从句中有两个并列部分，即 the equivalent circuit as derived here is only valid for normal operation at power frequencies 和 capacitance effects must be taken into account whenever the rate of change of voltage would give rise to appreciable capacitance currents，$I_C=CdV/dt$，用分号隔开。power frequency 意为"电网频率"；后一部分中 whenever 引导状语从句；take into account 意为 note or consider it；pay attention to，"对某事物加以考虑；对某事物加以注意"。

【10】An alternative, treating the transformer as a three or four-terminal network gives rise to a representation which is just as accurate and has some advantages for the circuit engineer who treats all devices as circuit elements with certain transfer properties.

这个长句的主干是 An alternative gives rise to a representation，treating the transformer as a three or four-terminal network 是动名词短语做主语 alternative 的同位语，对其进一步加以说明；which is just as accurate and has some advantages...定语从句修饰主句中的宾语 representation，其中又有一个由 who 引导的定语从句修饰从句中的介词宾语 circuit engineer；transfer properties 意为"传递性能"。

【11】The importance of understanding this preliminary transformer work cannot be overemphasized.

注意在翻译这个句子时，cannot be overemphasized 译为"怎么强调也不过分"，而不能译成"不能过于强调"。cannot 其意为"做不到，没有能力做到"，而 over 为"过分"之意，因此 cannot be overemphasized 就是"做不到过分强调"的意思，既然"做不到过分强调"，那不就是"怎么强调也不过分"吗？因此该句应该译为："了解这种简单变压器的工作情况的重要性怎么强调都不过分"。

2.1.4 Translation

变压器的工作原理

前面已提到，通过选择合适的匝数比，一次侧输入电压 V_1，可任意转换成所希望的二次侧开路电压 E_2。E_2 可用于产生负载电流，该电流的幅值和功率因数将由二次侧电路的阻抗决定。现在，我们要讨论一种滞后功率因数。二次侧电流及其总安匝 I_2N_2 将影响磁通，有一种对铁心产生去磁、减小 Φ_m 和 E_1 的趋向。因为一次侧漏阻抗压降如此之小，所以 E_1 的微小变化都将导致一次侧电流增加很大，从 I_0 增大至一个新值 $I_1=(V_1+E_1)/(R_1+jX_1)$。增加的一次侧电流和磁势近似平衡了全部二次侧磁势。这样的话，互感磁通只经历了很小的变化，并且实际上只需要与空载时相同的净磁势 I_0N_1。一次侧总磁势增加了 I_2N_2，它是平衡同量的二次侧磁势所必需的。在相量方程中，$\dot{I}_1N_1+\dot{I}_2N_2=\dot{I}_0N_1$，上式也可变换成 $\dot{I}_1N_1=\dot{I}_0N_1-\dot{I}_2N_2$。满载时，电流 I_0 只约占满载电流的 5%，因而 I_1 近似等于 I_2N_2/N_1。记住，近似等于 E_1I_1 的输入容量也就近似等于输出容量 E_2I_2。

此时若忽略电阻，其物理现象变化的图示可参照图 1（略）。一次侧电流已增大，随之与之成正比的一次侧漏磁通也增大。交链一次绕组的总磁通 $\Phi_p=\Phi_m+\Phi_1=\Phi_{11}$，没有变化，这是因为总反电动势（$E_1-N_1d\Phi_1/dt$）仍然与 V_1 相等且反向。然而此时却存在磁通的重新分配，由于

Φ_1 随 I_1 的增加而增加，互感磁通分量已经减小。尽管变化很小，但是如果没有互感磁通和电动势的变化来允许一次侧电流变化，那么二次侧的需求就无法满足。交链二次绕组的净磁通 Φ_s 由于 I_2 产生的二次侧漏磁通（其与 Φ_m 反向）的建立而被进一步削弱。尽管图中 Φ_m 和 Φ_2 是分开表示的，但它们在铁心中是一个合成量，该合成量在图示中的瞬时是向下的。这样，二次侧端电压降至 $V_2=-N_2 d\Phi_s dt$，它可被看成两个分量，即 $V_2=-N_2 d\Phi_m dt - N_2 d\Phi_2 dt$，或者相量形式 $\dot{V}_2 = \dot{E}_2 - jX_2\dot{I}_2$。与一次侧漏磁通一样，$\Phi_2$ 的作用也用一个大体为常数的漏电感 $N_2\Phi_2/i_2 = N_2^2 \Lambda_2$ 来表征。要注意的是，由于互感磁通的作用，一次侧漏磁通对于二次侧端电压的变化产生部分影响。这两种漏磁通紧密相关；例如，Φ_2 对 Φ_m 的去磁作用引起了一次侧的变化，从而导致了一次侧漏磁通的产生。

如前所述，通常用反相的电压降来表示漏感电动势。图 2（略）画出所有的磁通分量和各个电动势，其应与空载时的图相比较，后者画出了各电压降分量，并包括很小的电阻压降。

至此，前面分析的是一个滞后的功率因数。如果下面讨论一个足够低的超前功率因数，二次侧总磁通和互感磁通都会增加，从而使得二次侧端电压随负载增加而升高。在空载情形下，如果忽略电阻，Φ_p 幅值大小不变，因为它仍提供一个等于 V_1 的总的反电动势。尽管现在 Φ_p 是一次侧和二次侧磁势的共同作用产生的，但它实际上与 Φ_{11} 相同。互感磁通必须仍随负载变化而变化以改变 E_1，从而产生更大的一次侧电流。此时 E_1 的幅值已经增大，但由于 E_1 与 V_1 是相量合成，因此一次侧电流仍然是增大的。

从上述图中，还应得出两点：首先，为方便起见已假设匝数比为 1，这样可使 $E_1=E_2'$。其次，如果横轴像通常那样取的话，那么相量图是以 $\Phi_m=0$ 为零时间参数的，图中各物理量时间方向并不是瞬时的。在周期性交变中，有一次侧漏磁通为零的瞬时，也有二次侧漏磁通为零的瞬时，还有它们处于同一方向的瞬时。

等效电路

已经推出的变压器二次侧绕组端开路的等效电路，通过加上二次侧电阻和漏抗便可很容易扩展成二次侧负载时的等效电路。

实际中所有的变压器的匝数比都不等于 1，尽管有时使其为 1 也是为了使一个电路与另一个在相同电压下运行的电路实现电气隔离。为了分析 $N_1=N_2$ 时的情况，二次侧的反应可从一次侧来看，这种反应只有通过由二次侧的磁势产生磁场力来反映。我们从一次侧无法判断是 I_2 大、N_2 小，还是 I_2 小、N_2 大，正是电流和匝数的乘积在产生作用。因此，二次侧绕组可用任意多个在一次侧产生相同反应的不同的等效绕组和负载电路代替。将二次绕组变换成与一次侧具有相同匝数 N_1 的等效绕组是方便的。因此图 3（略）的电路会是适用的，尽管这需要将二次绕组的实际参数加以变换，即由于 N_2 变换成了 $N_2'=N_1$，因此 E_2, I_2, Z_2 就应变换成 E_2', I_2', Z_2'。

当 N_2 变换成 N_1，由于电动势与匝数成正比，所以 $E_2'=(N_1/N_2)E_2$，与 E_1 相等。

对于电流，由于对一次侧作用的安匝数必须保持不变，因此 $I_2'N_2'=I_2'N_1=I_2N_2$，即 $I_2'=(N_2/N_1)I_2$。

对于阻抗，由于二次侧电压 V 变成 $(N_1/N_2)V$，电流 I 变为 $(N_2/N_1)I$，因此阻抗值，包括负载阻抗必然变为 $V'/I' =(N_1/N_2)^2 V/I$。因此 $R_2'=(N_1/N_2)^2 R_2$，$X_2'=(N_1/N_2)^2 X_2$。

如果将一次侧匝数作为参考匝数，那么这种过程称为往一次侧的折算。

可以用一些方法来验证上述折算过程是否正确。

例如，折算后的二次绕组的铜耗必须与原二次绕组的铜耗相等，否则一次侧提供给其损耗的功率就变了。$I_2'^2 R_2'$ 必须等于 $I_2^2 R_2$，而 $(I_2 \cdot N_2/N_1)^2(R_2 \cdot N_1^2/N_2^2)$ 事实上确实简化成了 $I_2^2 R_2$。

类似地，与 $I_2^2 X_2$ 成比例的漏磁场的磁场储能 $\left(\frac{1}{2}LI^2\right)$，求出后验证与 $I_2'^2 X_2'$ 成正比。折算后的二次侧 kVA=$E_2'I_2'$=$E_2(N_1/N_2) \cdot I_2(N_2/N_1)$=$E_2 I_2$。

尽管看起来似乎不可理解，事实上这种论点是可靠的。实际上，如果将实际的二次绕组当真从铁心上移开，并用一个参数设计成 N_1, R_2', X_2', I_2' 的等效绕组和负载电路替换，在正常电网频率运行时，从一次侧两端无法判断二次侧的磁势、所需容量及铜耗与前面有何差别。

在选择折算基准时，无非是将一次侧与折算后的二次侧匝数设为相等，除此之外再没有什么更要紧的了。但有时将一次侧折算到二次侧倒是方便的，在这种情况下，如果所有下标"1"的量都变换成了下标"2"的量，那么很容易得到必需的折算系数，例如 $R_1' = R_1(N_2/N_1)^2$，值得注意的是，对于一台实际的变压器，$R_1' \approx R_2$，$X_1' \approx X_2$，同样地 $R_2' \approx R_1$，$X_2' \approx X_1$。

$N_1 \neq N_2$ 的通常情形时的等效电路，除了是为了考虑铁耗而引入了 r_m，且为了将 V_2' 折算回 V_2 而在二次侧两端引入了一理想的无损耗转换外，其他方面与图4（略）是一样的。在运用这种理想转换之前，内部电压和功率损耗已进行了计算。当在电路中选择了适当的参数时，在一、二次侧两端测得的变压器运行情况与在该电路相应端所测的情况是完全一致的。将 N_1 线圈和 N_2 线圈并排放置在一个铁心的两边，这一点与实际情况之间的差别仅仅是为了方便。当然，就变压器本身来说，两线圈是绕在同一铁心柱上的。

如果将激磁支路移至一次绕组端口，引起的误差很小，但一些不可思议的现象又会出现。例如，流过一次侧阻抗的电流不再是整个一次侧电流。由于 I_0 通常只是 I_1 的很小一部分，所有误差相当小。对一个具体问题可否允许有细微差别的回答取决于是否允许这种误差的存在。对于这种简化电路，一次侧和折算后二次侧阻抗可相加，得

$$R_{e1} = R_1 + R_2(N_1/N_2)^2 \text{ 和 } X_{e1} = X_1 + X_2(N_1/N_2)^2$$

需要指出的是，在此得到的等效电路仅适用于电网频率下的正常运行；一旦电压变化率产生相当大的电容电流 $I_C = CdV/dt$ 时，必须考虑电容效应。这对于高电压和频率超过 100Hz 的情形是很重要的。其次，图 5（略）即使是对于电网频率也并非唯一可行的等效电路。另一种形式是将变压器看成一个三端或四端网络，这样便产生一个准确的表达，它对于那些把所有装置看成是具有某种传递性能的电路元件的工程师来说是方便的。以此为分析基础的电路，会拥有一个既产生电压大小的变化，也产生相位移的匝比，其阻抗也会与绕组的阻抗不同。这种电路无法解释变压器内类似饱和效应等现象，所以对于掌握了解其内部的物理现象，还是图5（略）合适。

等效电路有两种端口形式：

（a）从一次侧看，为一个 U 形电路，其折合后的负载阻抗的端电压为 V_2'。

（b）从二次侧看，为一其值为 V_1，且伴有由 R_{e1} 和 X_{e1} 引起内压降的恒压源。在这种电路中有时可省略激磁支路，这样电路简化为一台产生恒值电压 E_1（实际上等于 V_1），并带有阻抗 $R + jX$（实际上等于 $R_{e1}+jX_e$）的发电机。在上述两种情况下，参数都可折算到二次绕组，这样可减小计算时间。其电阻和电抗值可通过两种简单的轻载试验获得。了解这种简单变压器的工作情况的重要性怎么强调都不过分。它与各种电机的原理均有关联，例如，图6（略）将是感应电机等效电路的形成基础。

Unit 2.2 D.C. Machine Construction and Principle

2.2.1 Text

D.C. Machine Construction

A D.C. machine is made up of two basic components:

The stator which is the stationary part of the machine. It consists of the following elements: a yoke inside a frame; excitation poles and winding; commutating poles (com-poles) and winding; end shield with ball or sliding bearings; brushes and brush holders; the terminal box.

The rotor which is the moving part of the machine. It is made up of a core mounted on the machine shaft. This core has uniformly—spaced slots into which the armature winding is fitted. A commutator, and often a fan, are also located on the machine shaft.

The frame is fixed to the floor by means of a bedplate and bolts. On low-power machines the frame and yoke are one and the same components, through which the magnetic flux produced by the excitation poles closes 【1】. The frame and the yoke are built of cast iron or cast steel or sometimes from welded steel plates.

In low-power and controlled rectifier-supplied machines the yoke is built up of thin (0.5—1mm) laminated iron sheets. The yoke is usually mounted inside a non-ferromagnetic frame (usually made of aluminium alloys, to keep down the weight). To either side of the frame there are bolted two end shields, which contain the ball or sliding bearings.

The (main) excitation poles are built from 0.5—1mm iron sheets held together by riveted bolts. The poles are fixed into the frame by means of bolts. They support the windings carrying the excitation current.

On the rotor side, at the end of the pole core is the so-called pole-shoe which is meant to facilitate a given distribution of the magnetic flux through the air gap. The winding is placed inside an insulated frame mounted on the core, and secured by the pole-shoe.

The excitation windings are made of insulated round or rectangular conductors, and are connected either in series or in parallel. The windings are linked in such a way that the magnetic flux of one pole crossing the air gap is directed from the pole-shoe towards the armature (north pole), which the flux of the next pole is directed from the armature to the pole-shoe (south pole).

The commutating poles, like the main poles, consist of a core ending in the pole-shoe and a winding wound round the core. They are located on the symmetry (neutral) axis between two main poles, and bolted on the yoke. Commutating poles are built either of cast-iron or iron sheets.

The windings of the commutating poles are also made from insulated round or rectangular conductors. They are connected either in series or in parallel and carry the machine's main current.

The rotor core is built of 0.5—1mm silicon-alloy sheets. The sheets are insulated from one another by a thin film of varnish or by an oxide coating, both some 0.03—0.05mm thick. The purpose is to ensure a reduction of the eddy currents which arise in the core when it rotates inside the magnetic field. These currents cause energy losses which turn into heat. In solid cores, these losses could become very high, reducing machine efficiency and producing intense heating.

The rotor core consists of a few packets of metal sheet. Redial or axial cooling ducts (8—10mm inside) are inserted between the packets to give better cooling. Pressure is exerted to both sides of the core by pressing devices foxed on to the shaft. The length of the rotor usually exceeds that of the poles by 2—5mm on either side—the effect being to minimize the variations in magnetic permeability caused by axial armature displacement. The periphery of the rotor is provided with teeth and slots into which the armature winding is inserted.

The rotor winding consists either of coils wound directly in the rotor slots by means of specially designed machines or coils already formed. The winding is carefully insulated, and it secured within the slots by means of wedges made of wood or other insulating material.

The winding overcharge are bent over and tied to one another with steel wire in order to resist the deformation which could be caused by the centrifugal force.

The coil-junctions of the rotor winding are connected to the commutator mounted on the armature shaft. The commutator is cylinder made of small copper segments insulated from one another, and also from the clamping elements by a layer of mica. The ends of the rotor coil are soldered to each segment.

On low-power machines, the commutator segments form a single unit, insulated from one another by means of a synthetic resin such as bakelite.

To link the armature winding to the fixed machine terminals, a set of carbon brushes slide on the commutator surface by means of brush holders. The brushes contact the commutator segments with a constant pressure ensured by a spring and lever. Clamps mounted on the end shields support the brush holders.

The brushes are connected electrically—with the odd-numbered brushes connected to one terminal of the machine and the even-numbered brushes to the other. The brushes are equally spaced round the periphery of the commutator—the number of rows of brushes being equal to the number of excitation poles.

D.C. Machine Principle

DC machines are characterized by their versatility. By means of various combinations of shunt, series, and separately excited field windings they can be designed to display a wide variety of volt-ampere or speed-torque characteristics for both dynamic and steady-state operation. Because of the ease with which they can be controlled, systems of DC machines are often used in applications requiring a wide range of motor speeds or precise control of motor output 【2】.

The essential features of a DC machine are shown schematically. The stator has salient poles and is excited by one or more field coils. The air-gap flux distribution created by the field winding is symmetrical about the centerline of the field poles. This is called the field axis or direct axis.

As we know, the AC voltage generated in each rotating armature coil is converted to DC in the external armature terminals by means of a rotating commutator and stationary brushes to which the armature leads are connected【3】. The commutator-brush combination forms a mechanical rectifier, resulting in a DC armature voltage as well as an armature mmf wave which is fixed in space. The brushes are located so that commutation occurs when the coil sides are in the neutral zone, midway between the field poles. The axis of the armature-mmf wave then is 90 electrical degrees from the axis of the field poles, i.e., in the quadrature axis. In the schematic representation the brushes are shown in quadrature axis because this is the position of the coils to which they are connected. The armature-mmf wave then is along the brush axis as shown. (The geometrical position of the brushes in an actual machine is approximately 90 electrical degrees from their position in the schematic diagram because of the shape of the end connections to the commutator.)

The magnetic torque and the speed voltage appearing at the brushes are independent of the spatial waveform of the flux distribution; for convenience we shall continue to assume a sinusoidal flux-density wave in the air gap. The torque can then be found from the magnetic field viewpoint.

The torque can be expressed in terms of the interaction of the direct-axis air-gap flux per pole Φ_d and the space-fundamental component F_{a1} of the armature-mmf wave. With the brushes in the quadrature axis the angle between these fields is 90 electrical degrees, and its sine equals unity. For a P pole machine

$$T=\frac{\pi}{2}\left(\frac{P}{2}\right)^2 \Phi_d F_{a1}$$

in which the minus sign gas been dropped because the positive direction of the torque can be determined from physical reasoning. The space fundamental F_{a1} of the sawtooth armature-mmf wave is $8/\pi^2$ times its peak. Substitution in above equation then gives

$$T=\frac{PC_a}{2\pi m}\Phi_d i_a = K_a \Phi_d i_a \quad N \cdot m$$

where i_a=current in external armature circuit

C_a=total number of conductors in armature winding

m=number of parallel paths through winding

and

$$K_a = \frac{PC_a}{2\pi m}$$

is a constant fixed by the design of the winding.

The rectified voltage generated in the armature has already been discussed before for an elementary single-coil armature. The effect of distributing the winding in several slots is shown in figure, in which each of the rectified sine waves is the voltage generated in one of the coils, commutation taking place at the moment when the coil sides are in the neutral zone. The generated voltage as observed from the brushes and is the sum of the rectified voltages of all the coils in series between brushes and is shown by the rippling line labeled e_a in figure. With a dozen or so commutator segments per pole, the ripple becomes very small and the average generated voltage observed from the brushes equals the sum of the average values of the rectified coil voltages. The rectified voltage e_a between brushes, known also as the speed voltage, is

$$e_a = \frac{PC_a}{2\pi m}\Phi_d\omega_m = K_a\Phi_d\omega_m$$

where K_a is the design constant. The rectified voltage of a distributed winding has the same average value as that of a concentrated coil. The difference is that the ripple is greatly reduced.

From the above equations, with all variable expressed in SI units,

$$e_a i_a = T\omega_m$$

This equation simply says that the instantaneous electric power associated with the speed voltage equals the instantaneous mechanical power associated with the magnetic torque, the direction of power flow being determined by whether the machine is acting as a motor or generator 【4】.

The direct-axis air-gap flux is produced by the combined mmf $\sum N_f i_f$ of the field windings, the flux-mmf characteristic being the magnetization curve for the particular iron geometry of the machine. In the magnetization curve, it is assumed that the armature-mmf wave is perpendicular to the field axis. It will be necessary to reexamine this assumption later in this chapter, where the effects of saturation are investigated more thoroughly. Because the armature e_{mf} is proportional to flux times speed, it is usually more convenient to express the magnetization curve in terms of the armature emf e_{a0} at a constant speed ω_{m0}. The voltage e_a for a given flux at any other speed ω_m is proportional to the speed, i.e.,

$$e_a = \frac{\omega_m}{\omega_{m0}}e_{a0}$$

Fig.1 (omitted) ure C shows the magnetization curve with only one field winding excited. This curve can easily be obtained by test methods, no knowledge of any design details being required.

Over a fairly wide range of excitation the reluctance of the iron is negligible compared with that of the air gap. In this region the flux is linearly proportional to the total mmf of the field windings, the constant of proportionality being the direct-axis air-gap permeance.

The outstanding advantages of DC machines arise from the wide variety of operating characteristics which can be obtained by selection of the method of excitation of the field windings. The field windings may be separately excited from an external DC source, or they may be self-excited; i.e., the machine may supply its own excitation. The method of excitation profoundly influences not only the steady-state characteristics, but also the dynamic behavior of the machine in control systems.

The connection diagram of a separately excited generator is given. The required field current is a very small fraction of the rated armature current. A small amount of power in the field circuit may control a relatively large amount of power in the armature circuit; i.e., the generator is a power amplifier. Separately excited generators are often used in feedback control systems when control of the armature voltage over a wide range is required. The field windings of self-excited generators may be supplied in three different ways. The field may be connected in series with the armature, resulting in a series generator. The field may be connected in shunt with the armature, resulting in a shunt generator, or the field may be in two sections, one of which is connected in series and the other in shunt with the armature, resulting in a compound generator. With self-excited generators residual magnetism must be present in the machine iron to get the self-excitation process started.

In the typical steady-state volt-ampere characteristics, the constant-speed prime movers are assumed. The relation between the steady-state generated e.m.f. E_a and the terminal voltage V_t is

$$V_t = E_a - I_a R_a$$

where I_a is the armature current output and R_a is the armature circuit resistance. In a generator, E_a is larger than V_t, and the electromagnetic torque T is a counter-torque opposing rotation.

The terminal voltage of a separately excited generator decreases slightly with increase in the load current, principally because of the voltage drop in the armature resistance. The field current of a series generator is the same as the load current, so that the air-gap flux and hence the voltage vary widely with load. As a consequence, series generators are normally connected so that the mmf of the series winding aids that of the shunt winding. The advantage is that through the action of the series winding the flux per-pole can increase with load, resulting in a voltage output which is nearly usually contains many turns of relatively small wire. The series winding, wound on the outside, consists of a few turns of comparatively heavy conductor because it must carry the full armature current of the machine. The voltage of both shunt and compound generators can be controlled over reasonable limits by means of rheostats in the shunt field.

Any of the methods of excitation used for generators can also be used for motors. In the typical steady-state speed-torque characteristics, it is assumed that the motor terminals are supplied from a constant-voltage source. In a motor the relation between the e.m.f. E_a generated in the armature and the terminal voltage V_t is

$$V_t = E_a + I_a R_a$$

where I_a is now the armature current input. The generated e.m.f. E_a is now smaller than the terminal voltage V_t, the armature current is in the opposite direction to that in a generator, and the electro-magnetic torque is in the direction to sustain rotation of the armature.

In shunt and separately excited motors the field flux is nearly constant. Consequently, increased torque must be accompanied by a very nearly proportional increase in armature current and hence by a small decrease in counter e.m.f. to allow this increased current through the small armature resistance. Since counter e.m.f. is determined by flux and sped, the speed must drop slightly. Like the squirrel-cage induction motor, the shunt motor is substantially a constant-speed motor having about 5 percent drop in speed from no load to full load. Starting torque and maximum torque are limited by the armature current that can be commutated successfully.

2.2.2　Specialized English Words

  end shield　（电机定子的）端盖
  terminal box　出线盒
  cast steel　铸钢
  film of vanish　漆膜
  wedge　槽楔
  shunt excited　并励

separately excited   他励
netural   中性的
quadrature axis   交轴
upper order harmonic component   高次谐波分量
concentrated coil   集中绕组
brush holder   握刷
cast iron   生铁
ferromagnetic   铁磁性的
oxide coating   氧化层
centrifugal force   离心力
series excited   串励
self excited   自励
direct axis   直轴
pace-fundamental component   空间基波分量
distributed winding   分布绕组
magnetization curve   磁化曲线

2.2.3 Notes

【1】On low-power machines the frame and yoke are one and the same components, through which the magnetic flux produced by the excitation poles closes.

句中介词短语 On low-power machines 做状语；句中主干为 the frame and yoke are one and the same components；限定性定语从句 through which the magnetic flux closes 修饰宾语 component，其中过去分词短语 produced by the excitation poles 做 magnetic flux 的后置定语。

【2】Because of the ease with which they can be controlled, systems of DC machines are often used in applications requiring a wide range of motor speeds or precise control of motor output.

句中 Because of 接名词 ease 形成介词短语表示原因，而 with which they can be controlled 是定语从句修饰 ease，with ease 意为 without difficulty，"容易的，无困难的"。若将该句扩展成原因状语从句，则为 Because they can be controlled with ease 或 Because they can be controlled easily。

【3】As we know, the AC voltage generated in each rotating armature coil is converted to DC in the external armature terminals by means of a rotating commutator and stationary brushes to which the armature leads are connected.

句中 As we know 是一个插入语，句子主干为 the AC voltage is converted to DC；generated in each rotating armature coil 是过去分词短语做主语 the AC voltage 的后置定语，译为"在每个旋转电枢线圈中产生的交流电压"；句子的状语有两部分：一为地点状语 in the external armature terminals，二为方式状语 by means of...，表明交流电压转换成直流的方法，其中包含一定语从句 to which the armature leads are connected 修饰 rotating commutator and stationary brushes，所修饰部分为介词 to 的对象，即 the armature leads are connected to the rotating commutator and stationary brushes。

【4】This equation simply says that the instantaneous electric power associated with the speed voltage equals the instantaneous mechanical power associated with the magnetic torque, the direction of power flow being determined by whether the machine is acting as a motor or generator.

句中 says 意为"表明，说明"；associated with 意为"与……相关"；整个句子是一个主从复合句，that 引导的从句 the instantaneous...with the magnetic torque 是主句中谓语的宾语，而后一部分 the direction of power flow being determined by whether the machine is acting as a motor or generator 是一独立分词短语，做主句的伴随状语，其中现在分词 being determined 的行为执行者为 the direction of power flow，即它便为独立分词短语中的主语，这种句型本文中较多。

2.2.4　Translation

直流电机的结构与工作原理

直流电机的结构

直流电机由两大主要部分组成。

定子：它是电机中的固定部分，由以下部件组成：机座内的磁轭、励磁极和励磁绕组、换向极和换向绕组、带滚珠轴承或滑动轴承的端盖、电刷及刷握、出线盒。

转子：它是电机中的转动部分，由装配在电机转轴上的铁心组成。铁心上有许多嵌放了绕组的均匀分布的槽，电机转轴上还装配有换向器和风扇。

机座通过底板和螺栓将整机固定在地面上。对于小功率电机，机座和磁轭常常合二为一，它是励磁磁极产生闭合磁通的通路。机座和磁轭由铸铁或铸钢铸成，有时或由钢板焊接而成。

在小功率和晶闸管整流电源供电的直流电机中，磁轭由 0.5～1mm 厚的薄铁片叠压而成，通常固定在非铁磁性机座（通常用铝合金制成以减轻质量）内。机座两边用螺栓固定着两个装有滚珠轴承或滑动轴承的端盖。

励磁磁极（主磁极）一般用 0.5～1mm 厚的铁片经叠压后用铆钉紧固而成。整个主极极身用螺杆固定在机座内，支撑通以励磁电流的绕组。

主极铁心靠近转子边一端的部分称为极靴，它的形状设计成便于形成所需的气隙磁密分布曲线。绕组嵌放于套在铁心上、并由极靴加以固定的绝缘框架内。

励磁绕组由圆形截面或矩形截面的绝缘导线绕制而成，它们既可以串联也可以并联的方式连接。绕组按这样的方式连接，使得一个极穿过气隙的磁通从极靴指向电枢（N 极），而下一个极的磁通由电枢指向极靴（S 极）。

换向极，和主极一样也由末端为极靴的铁心和套在上面的绕组构成。它们位于两个主极间的对称（中性）轴上，用螺杆固定在磁轭上。

换向极既可用生铁也可用铁片叠压而成。换向极绕组也是用圆形截面或矩形截面的绝缘导线绕制而成，它们也是既可用串联也可用并联的方式连接，其中通过的电流为主电流。

转子铁心用 0.5～1mm 厚的硅钢片叠压而成，叠片间用一层厚 0.03～0.05mm 的很薄的漆膜或氧化层相互绝缘。这种结构的目的在于保证转子铁心在磁场内旋转时铁心中产生的涡流有所减少。这些涡流产生转化成热能的能量损耗。在实心转子铁心中，这些损耗可能会非常大，从而降低了电机的效率并产生很大的热量。

转子铁心分成几段，为了加强冷却，在各段之间有许多辐射状或轴向的冷却通风孔，通过固定在转轴上的压力装置将风压施加在铁心两边。转子铁心长度通常每边超出磁极各 2～5mm，其作用是减小由轴向电枢位置的偏差而产生的磁导率的变化。转子圆周上有许多齿和内部嵌放有电枢绕组的槽。

转子绕组可由直接绕在经特殊设计的电机的转子槽中的线圈组成，或是由成形线圈组成。绕组必须经过妥善绝缘，并用由木头或者其他绝缘材料制成的槽楔固定在槽内。

为了防止绕组端部受到离心力作用而产生变形，必须将其弯曲并用钢丝把它们互相捆绑在一起。

转子绕组的线圈接头被连接到固定在电枢转轴上的换向器上。换向器是由许多小铜片做成的换向片组成的圆筒，换向片与换向片之间、换向片与压紧元件之间用一层云母垫片绝缘。每一转子线圈的出线端均焊接到一个换向片上。

对小功率直流电机，换向器的换向片形成一个独立整体，用类似酚醛塑料的合成树脂相互绝缘。

为了将电枢绕组引到电机定子上的出线端，用握刷固定的一组碳刷在换向器表面滑动。弹簧和刷杆保持电刷与换向片在弹簧提供的恒压下与换向片接触，固定在端盖上的夹紧机构支撑着握刷。

所有奇数号电刷与电机的一个出线端在电路上相连，而偶数号电刷与另一个出线端相连。电刷沿着换向器圆周等距离分布，电刷数与磁极数相等。

直流电机的工作原理

直流电机以其多功用性而形成了鲜明的特征。通过并励、串励和他励绕组的各种不同组合，直流电机可设计成在动态和稳态运行时呈现出宽广范围变化的伏安特性或速度—转矩特性。由于直流电机易于控制，因此该系统用于要求电机转速变化范围宽或能精确控制电机输出的场合。

直流电机的总貌如图所示。定子上有凸极，由一个或一个以上励磁线圈励磁。励磁绕组产生的气隙磁通以磁极中心线为轴线对称分布，这条轴线称为磁场轴线或直轴。

众所周知，每个旋转的电枢绕组中产生的交流电压，经由一与电枢连接的旋转的换向器和静止的电刷，在电枢绕组出线端转换成直流电压。换向器—电刷的组合构成机械整流器，它产生一直流电枢电压和一在空间固定的电枢磁势波形。电刷的放置应使换向线圈也处于磁极中性区，即两磁极之间。这样，电枢磁势波形的轴线与磁极轴线相差 90° 电角度，即位于交轴上。在示意图中，电刷位于交轴上，因为此处正是与其相连的线圈的位置。这样，如图所示电枢磁势的轴线也是沿着电刷轴线的（在实际电机中，电刷的几何位置大约偏移图例中所示位置 90° 电角度，这是因为元件的末端形状构成图示结果与换向器相连）。

电刷上的电磁转矩和速度电压与磁通分布的空间波形无关；为了方便起见，我们假设气隙中仍然是正弦磁密波，这样便可以从磁场分析着手求得转矩。

转矩可以用直轴的每极气隙磁通 Φ_d 和电枢磁势波的空间基波分量 F_{a1} 相互作用的结果来表示。电刷处于交轴时，磁场间的角度为 90° 电角度，其正弦值等于1，则对于一台 P 极电机

$$T = \frac{\pi}{2}\left(\frac{P}{2}\right)^2 \Phi_d F_{a1}$$

式中由于转矩的正方向可以根据物理概念的推断确定，因此负号已经去掉。电枢磁势锯齿波的空间基波 F_{a1} 是峰值的 $8/\pi^2$。上式变换后有

$$T = \frac{PC_a}{2\pi m} \Phi_d i_a = K_a \Phi_d i_a \quad \text{N·m}$$

式中，i_a 为电枢外部电路中的电流；C_a 为电枢绕组中的总导体数；m 为通过绕组的并联支路数。

$$K_a = \frac{PC_a}{2\pi m}$$

其为一个由绕组设计而确定的常数。

简单的单个线圈的电枢中的整流电压前面已经讨论过了。将绕组分散在几个槽中的效果可用图形表示，图中每一条整流的正弦波形是一个线圈产生的电压，换向线圈边处于磁场的中性区。从电刷端口观察到的电压是电刷间所有串联线圈中整流电压的总和，在图中由标以 e_a 的波线表示。当每极有大约十几个换向器片，波形线的波动变得非常小时，从电刷端观察到的平均电压等于线圈整流电压平均值之和。电刷间的整流电压 e_a 即速度电压，为

$$e_a = \frac{PC_a}{2\pi_m} \Phi_d \omega_m = K_a \Phi_d \omega_m$$

式中，K_a 为设计常数。分布绕组的整流电压与集中线圈有着相同的平均值，其差别只是分布绕组的波形脉动大大减小。

将上述几式中的所有变量用 SI 单位制表达，有

$$e_a i_a = T \omega_m$$

这个等式简单地说明与速度电压有关的瞬时功率等于与磁场转矩有关的瞬时机械功率，能量的流向取决于这台电机是电动机还是发电机。

直轴气隙磁通由励磁绕组的合成磁势 $\sum N_f i_f$ 产生，磁通—磁势曲线就是电机的具体铁磁材料的几何尺寸决定的磁化曲线。在磁化曲线中，因为电枢磁势的轴线与磁场轴线垂直，因此假定电枢磁势对直轴磁通不产生作用。这种假设有必要在后述部分加以验证，届时饱和效应会深入研究。因为电枢电势 e_{mf} 与磁通成正比，所以通常用恒定转速 ω_{m0} 下的电枢电势 e_{a0} 来表示磁化曲线更为方便。任意转速 ω_m 时，任一给定磁通下的电压 e_a 与转速成正比，即

$$e_a = \frac{\omega_m}{\omega_{m0}} e_{a0}$$

图 1（略）中表示只有一个励磁绕组的磁化曲线，这条曲线可以很容易通过实验方法得到，不需要任何设计步骤的知识。

在一个相当宽的励磁范围内，铁磁材料部分的磁阻与气隙磁阻相比可以忽略不计，在此范围内磁通与励磁绕组总磁势成线性比例，比例常数便是直轴气隙磁导率。

直流电机的突出优点是通过选择磁场绕组不同的励磁方法，可以获得变化范围很大的运行特性。励磁绕组可以由外部直流电源单独激磁，或者也可自励，即电机提供自身的励磁。励磁方法不仅极大地影响控制系统中电机的静态特性，而且影响其动态运行。

他励发电机的连接图已经给出，所需励磁电流是额定电枢电流的很小一部分。励磁电路中很小数量的功率可以控制电枢电路中相对很大数量的功率，也就是说发电机是一种功率放大器。当需要在很大范围内控制电枢电压时，他励发电机常常用于反馈控制系统中。自励发电机的励磁绕组可以有三种不同的供电方式。励磁绕组可以与电枢串联起来，这便形成了串励发电机；励磁绕组可以与电枢并联在一起，这便形成了并励发电机；或者励磁绕组分成两部分，其中一部分与电枢串联，另一部分与电枢并联，这便形成了复励发电机。为了引起自励过程，在自励发电机中必须存在剩磁。

在典型的静态伏安特性中，假定原动机恒速运行，稳态电势 E_a 和端电压 V_t 的关系为

$$V_t=E_a-I_aR_a$$

式中，I_a 为电枢输出电流，R_a 为电枢回路电阻。在发电机中，E_a 比 V_t 大，电磁转矩 T 是阻转矩。

他励发电机的端电压随着负载电流的增加稍有降低，这主要是由于电枢电阻上的压降。串励发电机中的励磁电流与负载电流相同，这样，气隙磁通和电压随负载变化很大，因此很少采用串励发电机。并励发电机的电压随负载增加会有所下降，但在许多应用场合，这并不妨碍使用。复励发电机的连接通常是使串励绕组的磁势与并励绕组磁势相加，其优点是通过串励绕组的作用，每极磁通会随着负载的增加而增加，从而产生一个随负载增加近似为常数的输出电压。通常，并励绕组匝数多，导线细；而绕在外部的串励绕组由于它必须承载电机的整个电枢电流，所以其构成的导线相对较粗。不论是并励还是复励发电机的电压，都可以借助并励磁场中的变阻器在适度的范围内得到调节。

任何用于发电机的励磁方法都可用于电动机。在电动机典型的静态转速—转矩特性中，假设电动机两端由一个恒压源供电。在电动机电枢中感应的电势与端电压 V_t 间的关系为

$$V_t=E_a+I_aR_a$$

式中，I_a 为输入的电枢电流。电势 E_a 此时比端电压小，电枢电流与发电机中的方向相反，且电磁转矩与电枢旋转方向相同。

在并励和他励电动机中，磁场磁通近似为常数，因此转矩的增加必须要求电枢电流近似成比例增大，同时为允许增大的电流通过小的电枢电阻，要求反电势稍有减少。由于反电势决定于磁通和转速，因此，转速必须稍稍降低。与鼠笼式感应电动机类似，并励电动机实际上是一种从空载到满载速降仅约为 5% 的恒速电动机。启动转矩和最大转矩受到能成功换向的电枢电流的限制。

Unit 2.3　Three-Phase Induction Machine Principle

2.3.1　Text

The induction machine may be considered a transformer not only when the rotor is at standstill but also when the rotor is running. In this case the machine represents a universal transformer, i.e., a transformer in which not only the voltages, currents and number of phases undergo transformation, but also the frequency and the kind of energy 【1】. As a result, having written the e.m.f. equations of an induction machine and solved them for current, we obtain essentially the same equivalent circuits as for a transformer 【2】. Being descriptive and simple, the equivalent circuits are very useful for solving a number of problems, including that of circle diagram construction. Here we must agree to the following:

(1) That only the first harmonic of the variables (voltages, currents, etc.) are taken into account since they embrace and determine the main group of processes occurring in an induction machine;

(2) Consider the processes occurring in the rotor with the latter running at any speed irrespective of the cause which makes the rotor run, in order to make the analysis as generalized as possible.

Suppose the stator of an induction machine is connected to a circuit with a given voltage U_1 and a constant frequency f_1. The main flux Φ_m, rotating with a speed $n_1 = f_1 : p$, creates the fundamental e.m.f. E in the stator winding. In the same winding the primary leakage flux $\Phi_{\sigma 1}$ induces the leakage e.m.f. $\dot{E}_{\sigma 1} = -j\dot{I}_1 x_1$. If, in addition, we take into account the stator winding active resistance r_1, it may be seen that the stator of an induction machine when the rotor is running has the same e.m.f.s as in a machine with a braked rotor; in accordance with this, the e.m.f. equations written identically in both cases, namely

$$\dot{U}_1 = -\dot{E}_1 + \dot{I}_1 Z_1 \text{【3】}$$

According to the accepted conditions, the rotor may run either in the same direction as the field or in the opposite direction. In the first case we shall assume the rotor speed n to be positive, in the second case negative. Let us consider what occurs in the rotor, supposing the rotor circuit to be, so far, open; for this purpose the sliding contact of the rheostat is shifted to position 0 【4】.

(1) Frequency of the e.m.f. induced in the rotor winding

When a rotor runs at a speed n in a magnetic field rotating at a speed n_1, everything takes place as if the rotor were stationary and flux Φ_m rotated in respect to the rotor with a speed

$$n_2 = n_1 - n.$$

The frequency of the e.m.f. induced in the rotor winding is therefore equal to:

$$f_2 = pn_2 = p(n_1 - n) = n_1 p \frac{n_1 - n}{n_1} = f_1 s$$

where f_1 is the supply circuit frequency and s is the slip. We see that for a given frequency the e.m.f.

in the rotor changes directly in proportion to the slip. For shortness, the frequency f_2 is referred to as the slip frequency. In accordance with the slip variation range from $s=-\infty$ when generating to $s=+\infty$, when under electromagnetic brake operating conditions ($n<0$), the frequency f_2 also varies from $f_2=-\infty$ to $f_2=+\infty$ [5]. The signs "plus" and "minus" at a frequency f_2 have an arbitrary meaning, determining the change in sign of the induced e.m.f. when passing from one condition of operation to another.

(2) Rotor e.m.f.

According to the general expression for the rotor e.m.f. we have

$$E_{2s}=4.44f_2\omega_2 k_{\omega 2}\Phi_m=4.44f_1 s\omega_2 k_{\omega 2}\Phi_m=E_2 s$$

or, if the rotor winding is referred to the stator winding.

$$E'_{2s}=E'_2 s$$

That is to say, for a given main flux Φ_m, the e.m.f. induced in the rotor during rotation is equal to the e.m.f. E_2 when the rotor is at standstill multiplied by the slip [6]. If, for example, at $n=0$ with the rotor circuit open we obtain across the rings a voltage $U_2=E_2=600V$, by gradually increasing the speed of the rotor in the direction of the field from $n=0$ to $n=n_1$, we obtain a linear variation of E_{2s} from $E_{2s}=600V$ to $E_{2s}=0$; when $n>n_1$ e.m.f. E_{2s} begins to rise having a negative value, i.e., its phase changes in respect to the initial phase by 180° [7].

(3) Rotor winding resistance

Suppose the rotor circuit is closed through some additional resistance; for this we shift the sliding contact of the rheostat inserted into the rotor circuit from position 1 to position 3. Assume this to be an active resistance, since it corresponds more closely to the actual service conditions of an induction machine with slip rings. The active resistance of the rotor circuit will then be $R_2=r_2+r_{add}$, where r_2 is the active resistance of the rotor proper and r_{add} is the additional resistance.

If the phenomena of skin-effect in the rotor winding conductors and the change in winding active resistance due to change in temperature are disregarded, it may be assumed that

$$R_2=r_2+r_{add}=const$$

or, referring these resistances to the stator winding,

$$R'_2=r'_2=r'_{add}=const$$

The leakage reactance of the rotor at standstill $x_2=2\pi f_1 L_{\sigma 2}$, where $L_{\sigma 2}$ is the inductances due to the secondary leakage flux. Since the leakage fluxes pass mainly through air, $L_{\sigma 2}\approx const$. Consequently, the reactance of the rotor during rotation is

$$x_{2s}=2\pi f_2 L_{\sigma 2}=2\pi f_1 s L_{\sigma 2}=x_2 s$$

or, when referred to the stator winding, $x'_{2s}=x'_2 s$.

That is, the reactance of a rotor winding during rotation is equal to the reactance of the rotor when at standstill multiplied by the slip.

If the rotor circuit is closed, through it flows the current I_2 which creates the leakage flux $\Phi_{\sigma 2}$ and meets a resistance r_2. Accordingly, in the rotor winding exist an e.m.f. $\dot{E}_{2s}=\dot{E}_2 s$ produced by the main flux Φ_m and a leakage e.m.f. $\dot{E}_{\sigma 2}=-j\dot{I}_2 x_{2s}=-j\dot{I}_2 x_2 s$. Then according to Kirchhoff's second law,

$$\dot{E}_{2s}+\dot{E}_{\sigma 2}=\dot{E}_2 s-j\dot{I}_2 x_2 s=\dot{I}_2 r_2$$

or
$$\dot{E}_{2s} = \dot{I}_2 Z_{2s}$$

where $Z_{2s} = r_2 + jx_2 s$ is the impedance of the real rotor.

Consequently,
$$\dot{I} = \frac{\dot{E}_{2s}}{Z_{2s}} = \frac{\dot{E}_2 s}{r_2 + jx_2 s}$$

and
$$I_2 = \frac{E_2 s}{\sqrt{r_2^2 + x_2^2 s^2}}$$

If the rotor winding is referred to the stator winding; then
$$\dot{E}'_{2s} = \dot{I}'_2 Z'_{2s}$$

where $Z'_{2s} = r'_2 + jx'_2 s$ is the impedance of the referred rotor.

Hence,
$$\dot{I}'_2 = \frac{\dot{E}'_2 s}{r'_2 + jx'_2 s}$$

and
$$\dot{I}'_2 = \frac{\dot{E}'_2 s}{\sqrt{r_2'^2 + (x'_2)^2 s^2}}$$

In flowing through the rotor winding, the current I_2 creates an m.m.f. F_2 rotating in respect to the rotor with a speed n_2 which corresponds to the rotor-current frequency f_2. Furthermore, the rotor proper runs with speed n. Consequently, the rotor m.m.f. F_2 runs in respect to some fixed point in space and, also, therefore, in respect to the stator with a speed $n_2 + n$. But

$$n_2 = \frac{f_2}{p} = \frac{f_1 s}{p} = n_1 s = n_1 \frac{n_1 - n}{n_1} = n_1 - n$$

Thus,
$$n_2 + n = n_1 - n + n = n_1$$

That is, the rotor m.m.f. rotates in space always (i.e., irrespective of operating conditions) with the same speed and in the same direction as the stator m.m.f.s of an induction machine, we have agreed to consider only their first harmonics.

(1) M.M.F. equation

Since in an induction machine the stator and rotor m.m.f.s F_1 and F_2 both rotate in space with the same speed and in the same direction, it may be assumed that they are stationary in respect to each other and, consequently, create a common rotating sine wave of m.m.f. F_m[8]. The m.m.f. sine wave in this case must be shifted in space in respect to the m.m.f. F_1 by an angle that makes the resultant m.m.f. F_m sufficient for setting up the e.m.f. equilibrium of the main magnetic flux Φ_m necessary according to the conditions.

Thus,
$$\dot{F}_1 + \dot{F}_2 = \dot{F}_m$$

Substituting here for the m.m.f. values of equation we obtain
$$m_1 \omega_1 k_{\omega 1} \dot{I}_1 + m_2 \omega_2 k_{\omega 2} \dot{I}_2 = m_1 \omega_1 k_{\omega 1} \dot{I}_m$$

or
$$\dot{I}_1 + \frac{m_2 \omega_2 k_{\omega 2}}{m_1 \omega_1 k_{\omega 1}} \dot{I}_2 = \dot{I}_1 + \frac{1}{k_1} \dot{I}_2 = \dot{I}_m$$

We see that the m.m.f. equations of an induction machine with the rotor in motion are similar to the m.m.f. equations of an induction machine for $n=0$. Accordingly, the m.m.f. picture of an induction machine with the rotor in motion is similar to the m.m.f. picture of an induction machine for $n=0$, with the only difference that the speed in space n_1 of m.m.f. F_2 is made up of the speed $n_2=n_1-n$ of m.m.f. F_2 in respect to the rotor and the speed n of the rotor proper.

The conclusion reached above is true for any operating conditions of the induction machine. As long as it operates as a motor, the speed n is positive, i.e., the rotor m.m.f. rotates in the same direction as the rotor. When the machine operates as a generator, the speed n is negative, i.e., the rotor m.m.f. runs in a direction opposite to the rotor rotation. The same takes place when the machine operates as an electromagnetic brake.

（2）Vector diagrams of an induction machine as a particular case of a transformer

The mutual induction flux $\dot{\Phi}_m$ is defined by the value of the magnetizing current $\dot{I}_m$ which is equal to the vector sum of the current $\dot{I}_1$ and the referred current $\dot{I}_2'$. Neglecting the losses in the steel, it may be assumed that the flux $\Phi=\Phi_m$ is phase with the current $\dot{I}_m$. Let us introduce into the discussion the fictitious fluxes $\dot{\Phi}_1$ and $\dot{\Phi}_2$ coinciding in phase with the currents $\dot{I}_1$ and $\dot{I}_2$ and proportional to these currents so that

$$\frac{\dot{\Phi}_m}{\dot{I}_m} = \frac{\dot{\Phi}_1}{\dot{I}_1} = \frac{\dot{\Phi}_2}{\dot{I}_2}$$

In this case, it may be assumed that the flux $\dot{\Phi}_m$ is equal to the vector sum of the fluxes $\dot{\Phi}_1$ and $\dot{\Phi}_2$, that is

$$\dot{\Phi}_m = \dot{\Phi}_1 + \dot{\Phi}_2$$

2.3.2 Specialized English Words

general expression　　通式，公式
be referred to　　折算到，折合到

2.3.3 Notes

【1】In this case the machine represents a universal transformer, i.e., a transformer in which not only the voltages, currents and number of phases undergo transformation, but also the frequency and the kind of energy.

句中 represents 意为"相当于"，universal 意为"广义的，一般意义"；In this case 为介词短语，做全句的状语；a universal transformer 和 a transformer in which 为同位语，只是前一 transformer 由形容词 universal 做定语，后一个 transformer 由 in transformer 做定语；而该定语从句又是一个由 not only...but also...连接的并列复合句，只是后一子句省略了前一句相同的谓语 undergo，即整个定语从句实际上是 not only the voltages, currents and number of phases undergo transformation, but also the frequency and the kind of energy undergo transformation。

【2】As a result, having written the e.m.f. equations of an induction machine and solved them for current, we obtain essentially the same equivalent circuits as for a transformer.

句中 As a result 意为"因此",其结构;having written the e.m.f. equations of an induction machine and solved them...是分词的完成形式,做句子 we obtain essentially the same equivalent circuits as for a transformer.的状语,由于是完成时形式,因此表示其动作在句子谓语动词之前,说明"先列写出电势方程并求解电流,然后才得到类似变压器的等值电路"。to solve...for...意为"解……求……"。

【3】If, in addition, we take into account the stator winding active resistance r_1, it may be seen that the stator of an induction machine when the rotor is running has the same e.m.f.s as in a machine with a braked rotor; in accordance with this, the e.m.f. equations written identically in both cases, namely $\dot{U}_1 = -\dot{E}_1 + \dot{I}_1 Z_1$.

首先,in addition 是插入语,实际上此处应为 in addition, if we take into account...,即介词短语 in addition 放在句首做状语,但科技英语和专业英语中,往往将此短语放在句中,做插入语,前后用","分开,应习惯这种用法。其次,it 是主句的形式主语,真正的主语是 that 引导的主语从句,从句的基本结构是 the stator of an induction machine...has the same e.m.f.s as...,其中的 when the rotor is running 是 an induction machine 的定语从句,全句至此意思未完,因此用插入语 in accordance with this 引出一独立分词短语 the e.m.f. equations written identically in both cases 进一步将意思表达完整。in accordance with 意为"据此";namely 意为 that is to say,"即"。

【4】Let us consider what occurs in the rotor, supposing the rotor circuit to be, so far, open; for this purpose the sliding contact of the rheostat is shifted to position 0.

首先,应该看到这是一个复合句,用";"将两个从句分开,相当于用 and 将两个从句连接起来。supposing the rotor circuit to be, so far, open 是独立分词短语,其中 so far 是插入语,遇到这种情况,翻译时可以先不看它,这样短语的结构和意思是清楚的,supposing the rotor circuit to be open,再把插入语 so far 加上去,变成 supposing the rotor circuit to be open so far。整个短语的意思就清楚了,不会被中间的插入语弄乱了。在科技英语和专业英语的阅读和翻译中,应学会这种处理方法。so far 即"迄今为止"。

【5】In accordance with the slip variation range from $s=-\infty$ when generating to $s=+\infty$, when under electromagnetic brake operating conditions($n<0$), the frequency f_2 also varies from $f_2=-\infty$ to $f_2=+\infty$.

句中 In accordance with 意为"根据……",介词短语做状语。其中介词 with 又接名词 the slip variation range 做它的宾语,又形成一个介词短语结构。这个名词后又有一个介词短语 from $s=-\infty$ when generating to $s=+\infty$,做其后置定语,表示该 range 的范围。而由 when 分别引导的分词短语 when generating 和 when under electromagnetic brake operating conditions 分别做 $s=-\infty$ 和 $s=+\infty$ 的后置定语。

【6】For a given main flux Φ_m, the e.m.f. induced in the rotor during rotation is equal to the e.m.f. E_2 when the rotor is at standstill multiplied by the slip.

这句的主干是 the e.m.f. induced in the rotor...is equal to the e.m.f. E_2...multiplied by the slip。is equal to 即"等于",multiplied by the slip 即过去分词做 the e.m.f. E_2 的后置定语表示"乘以转差率"。

【7】If, for example, at $n=0$ with the rotor circuit open we obtain across the rings a voltage $U_2=E_2=600V$, by gradually increasing the speed of the rotor in the direction of the field from $n=0$ to

$n=n_1$, we obtain a linear variation of E_{2s} from E_{2s}=600V to E_{2s}=0; when $n> n_1$ e.m.f. E_{2s} begins to rise having a negative value, i.e., its phase changes in respect to the initial phase by 180°.

这是一个难长句。首先，整句是一个用";"隔开的两个从句组成的并列复合句。前一个并列从句又是一个主从复合句，其中主句是 we obtain a linear variation of E_{2s} from E_{2s}=600V to E_{2s}=0；其中 from E_{2s}=600V to E_{2s}=0 是介词短语做 a linear variation 的后置定语。by gradually increasing the speed of the rotor in the direction of the field from n=0 to $n= n_1$，是介词短语做该句的状语；If, for example, at n=0 with the rotor circuit open we obtain across the rings a voltage U_2=E_2=600V 则是前一个子句的条件状语从句，其中 for example 是插入语；at n=0, with the rotor circuit open 是两个介词短语做状语（也是插入语的形式），其实该句的句干为 If we obtain a voltage U_2=E_2=600V，across the rings 也是介词短语，在句中做地点状语；后一个并列从句比较简单，when $n> n_1$ 为状语。句干为 e.m.f. E_{2s} begins to rise having a negative value，i.e.引出的句子也可以看成是它的并列从句。

【8】Since in an induction machine the stator and rotor m.m.f.s F_1 and F_2 both rotate in space with the same speed and in the same direction, it may be assumed that they are stationary in respect to each other and, consequently, create a common rotating sine wave of m.m.f. F_m.

这是一个主从复合句，Since 引导的是一个状语从句，it may be assumed that...是主句。主句中 it 是形式主语，that 引导的才是真正的主语。主语从句中的主语是 they，它的谓语有两个动词 are stationary 和 create。

2.3.4　Translation

三相感应电机的工作原理

感应电机不仅在转子静止时，而且在转子旋转时也可以看作一变压器。在这种情况下，感应电机相当于一台广义上的变压器，这种变压器不仅其电压、电流和相数经历了转换，而且其频率和能量形式也经历了转换。因此，列写出感应电机的电势平衡方程并经其求解电流后，基本上获得了与变压器相同的等值电路。由于该等值电路表述清楚、结构简单，因此对求解包括绘制圆图在内的许多问题是很有用的。在此，必须达成以下共识：

（1）由于对诸如电压、电流等这些变量来说，其基波才参与和决定感应电机运行中各物理过程，因此，以下分析中仅考虑它们的作用。

（2）为了使分析尽可能更具普遍性，在考虑处于任一速度时转子中出现的各物理过程时，不涉及令其旋转的动因。

假设感应电机定子接于一给定电压为 U_1，恒定频率为 f_1 的电网。以 $n_1 = f_1/p$ 的转速旋转的主磁通 Φ_m 在定子绕组中产生基波电势 E_1。在同一绕组中，原边漏磁通 $\Phi_{\sigma 1}$ 感应漏电势 $\dot{E}_{\sigma 1}$ = $-\mathrm{j}\dot{I}_1 x_1$。此外，如果计及定子绕组纯电阻 r_1，可以看出，由于感应电机的电阻在转子旋转时所具有的电势与转子堵转时的电势相同，因此，两种情况下电势平衡方程的书写形式是相同的，即

$$\dot{U}_1 = -\dot{E}_1 + \dot{I}_1 Z_1$$

根据转子上承受转矩的具体状况，它可能与磁场同方向旋转，也可能与其反方向旋转。在前一种情况下，假定转子转速 n 为正，后一种情况下为负。假定转子目前仍处于开路状态

（为此目的，电阻箱的滑动触头应移至位置0），下面看看转子中将出现什么情况。

(1) 转子绕组感应电势的频率

转子在以转速 n_1 旋转的磁场中以转速 n 旋转时，期间所发生的一切物理过程犹如转子静止，而磁通 Φ_m 相对转子以

$$n_2 = n_1 - n$$

的转速旋转时是一样的。

因此，在转子绕组中感应电势的频率为

$$f_2 = pn_2 = p(n_1 - n) = n_1 p(n_1 - n)/n_1 = f_1 s$$

式中，f_1 为供电电源频率，s 为转差率。可见，频率一定，转子中感应电势与转差率成正比。为简便起见，频率 f_2 便视为转差频率。

根据转差率变化范围从 $s = -\infty$（发电状态时）到 $s = +\infty$（$n < 0$ 的电磁制动运行状态）可知，频率 f_2 的变化范围也是从 $f_2 = -\infty$ 到 $f_2 = +\infty$。频率中的符号"+"和"-"具有人为的特定含义，用来判定从一种运行状态到另一种运行状态时感应电势符号的变化。

(2) 转子电动势

根据转子感应电动势的计算公式，得

$$E_{2s} = 4.44 f_2 \omega_2 k_{\omega 2} \Phi_m = 4.44 f_1 s \omega_2 k_{\omega 2} \Phi_m = E_2 s$$

若转子绕组折合到定子边，则

$$E'_{2s} = E'_2 s$$

也就是说，对于某些给定的主磁通 Φ_m，旋转时在转子感应的电动势等于转子静止时的电动势 E_2 乘以转差率 s。例如，若 $n = 0$，且转子电路开路，转子滑环两端电压为 $U_2 = E_2 = 600\text{V}$，这时沿磁场转向从 $n = 0$ 到 $n = n_1$ 逐步增大转子速度，便将得到从 $E_{2s} = 600\text{V}$ 到 $E_{2s} = 0$ 的一个线性变化的 E_{2s}；当 $n > n_1$ 时，电动势 E_{2s} 的绝对值开始增大，但其值为负，即其相对于初相位变化了 $180°$。

(3) 转子绕组电阻

假设转子电路经一外加电阻闭合，为此，将串接在转子电路中的电阻箱的滑动触头由位置 1 移至位置 3。假设其为一纯电阻，这样便更接近带滑环的感应电机的实际工作状况。那么，转子电路的纯电阻将是 $R_2 = r_2 + r_{add}$，式中，r_2 为转子自身电阻，r_{add} 为外加电阻。

若转子绕组中导体的趋肤现象和绕组的纯电阻由于温度变化而变化的量忽略不计，便可认为

$$R_2 = r_2 + r_{add} = 常数$$

若将此折合到定子绕组，则

$$R'_2 = r'_2 + r'_{add} = 常数$$

转子静止时的漏电抗 $x_2 = 2\pi f_1 L_{\sigma 2}$，式中 $L_{\sigma 2}$ 为副边漏磁通产生的电感。由于漏磁通主要途径在空气，因此 $L_{\sigma 2}$ 近似为常数。因此，转子旋转时的电抗为

$$x_{2s} = 2\pi f_2 L_{\sigma 2} = 2\pi f_1 s L_{\sigma 2} = x_2 s$$

当折合到定子绕组时

$$x'_{2s} = x'_2 s$$

也就是说，旋转时转子绕组的电抗等于静止时转子电抗乘以转差率 s。

若转子电路闭合，则其间流过的电流 I_2 产生漏磁通 $\Phi_{\sigma 2}$，并遇到一电阻 r_2。因此，在转子绕组中便存在一个由主磁通 Φ_m 产生的电势 $E'_{2s} = E'_2 s$ 和一个漏电势 $\dot{E}_{\sigma 2} = -\text{j} \dot{I}_2 x_{2s} = -\text{j} \dot{I}_2 x_2 s$。这

样，根据基尔霍夫第二定律，便有
$$\dot{E}_{2s} + \dot{E}_{\sigma 2} = \dot{E}_{2}s - j\dot{I}_{2}x_{2}s = \dot{I}_{2}r_{2}$$
即
$$\dot{E}_{2s} = \dot{I}_{2}Z_{2s}$$
式中，$Z_{2s} = r_2 + jx_2 s$ 为未经折算的转子阻抗。

因此
$$\dot{I} = \frac{\dot{E}_{2s}}{Z_{2s}} = \frac{\dot{E}_{2}s}{r_2 + jx_2 s}$$

及
$$I_2 = \frac{E_2 s}{\sqrt{r_2^2 + x_2^2 s^2}}$$

若转子绕组折算到定子边，则
$$\dot{E}'_{2s} = \dot{I}'_{2}Z'_{2s}$$

式中，$Z'_{2s} = r'_2 + jx'_2 s$ 为经过折算的转子阻抗。

因此
$$\dot{I}'_2 = \frac{\dot{E}'_{2}s}{r'_2 + jx'_2 s}$$

及
$$\dot{I}'_2 = \frac{\dot{E}'_{2}s}{\sqrt{r'^2_2 + (x'_2)^2 s^2}}$$

转子电流 I_2 在转子绕组中流过时，产生一相对于转子以 n_2 速度旋转的磁势 F_2，该转速 n_2 由转子电流频率 f_2 决定。此外，转子本身以转速 n 旋转。因此，转子磁势 F_2 相对空间的某一静止物，也即相对于定子以 $n_2 + n$ 的速度旋转。而

$$n_2 = \frac{f_2}{p} = \frac{f_1 s}{p} = n_1 s = n_1 \frac{n_1 - n}{n_1} = n_1 - n$$

因此
$$n_2 + n = n_1 - n + n = n_1$$

也就是说，转子磁势在空间总是（即与运行状况无关）以与定子磁势相同的转速和按相同方向旋转。如果再次回忆前面所做的假设，便可知在谈及感应电机磁势时，已达成过共识，即仅考虑其基波。

（1）磁势方程

在感应电机中，由于定、转子磁势 F_1 和 F_2 均以相同转速按同一方向在空间旋转，因此可以认为它们是相对静止的，且产生一合成正弦波磁势 F_m。此时该正弦波磁势必定在空间相对于磁势 F_1 偏移一角度，以使合成磁势 F_m 能建立与运行时所必需的主磁通 Φ_m 相应的电势。因此
$$\dot{F}_1 + \dot{F}_2 = \dot{F}_m$$

用磁势幅值表达式代入上式，得
$$m_1 \omega_1 k_{\omega 1} \dot{I}_1 + m_2 \omega_2 k_{\omega 2} \dot{I}_2 = m_1 \omega_1 k_{\omega 1} \dot{I}_m$$

或
$$\dot{I}_1 + \frac{m_2 \omega_2 k_{\omega 2}}{m_1 \omega_1 k_{\omega 1}} \dot{I}_2 = \dot{I}_1 + \frac{1}{k_i} \dot{I}_2 = \dot{I}_m$$

可以看出，转子旋转时感应电机的磁势方程与其在 $n = 0$ 时的磁势方程是相似的。因此，感应电机转子旋转时的磁势图也与 $n = 0$ 的磁势图相类似，仅有一处不同的是，磁势 F_2 在空间的转速 n_1 是由磁势 F_2 相对于转子的速度 $n_2 = n_1 - n$ 和转子本身的速度 n 相叠加而成的。

以上分析所得结论对感应电机任何运行状况都是正确的。只要作为电动机运行，转速 n_2 便为正值，即转子磁势与转子同方向旋转。作为发电机运行时，n_2 为负值，即转子磁势与转

子转向相反。当电机作为电磁制动运行时,也是如此。

(2) 作为变压器特殊运行状态的感应电机的相量图

互感磁通 $\dot{\Phi}_m$ 由激磁电流 $\dot{I}_m$ 决定,而 $\dot{I}_m$ 则是 $\dot{I}_1$ 和折算后的 $\dot{I}_2'$ 的相量和,忽略铁损耗,可以认为主磁通 $\dot{\Phi}_m$ 与激磁电流 $\dot{I}_m$ 同相位。在此不妨引入假想磁通 $\dot{\Phi}_1$ 和 $\dot{\Phi}_2$,且 $\dot{\Phi}_1$ 和 $\dot{\Phi}_2$ 分别与 $\dot{I}_1$ 和 $\dot{I}_2$ 同相且大小成正比,因此

$$\frac{\dot{\Phi}_m}{\dot{I}_m} = \frac{\dot{\Phi}_1}{\dot{I}_1} = \frac{\dot{\Phi}_2}{\dot{I}_2}$$

在此情况下便可认为磁通 $\dot{\Phi}_m$ 等于磁通 $\dot{\Phi}_1$ 和 $\dot{\Phi}_2$ 的相量和

$$\dot{\Phi}_m = \dot{\Phi}_1 + \dot{\Phi}_2$$

Unit 2.4 Operation Analysis of Three-Phase Synchronous Generator

2.4.1 Text

The voltage diagram is of very great importance for analyzing working conditions in a synchronous machine. It is possible to obtain from the voltage diagram the percent variation of the synchronous generator voltage, the voltage increase with a drop in load and drop voltage for the transition from operation on no-load to operation on-load. The solution of these problems is of great importance: (1) for initial machine design when the necessary excitation current values are to be determined under various operating conditions and (2) when testing a finished machine to decide whether the machine conforms to given technical specifications. By using a voltage diagram, it is also possible to determine the operating conditions of a machine without actually applying the load, something which becomes especially difficult when the machine is of large rating.

The voltage diagrams make it possible to obtain the fundamental performance characteristics of a machine by means of calculation. Finally, the voltage diagram allows to determine the power angle θ between the e.m.f. produced by the excitation field and the voltage across the terminals. Angle θ plays a very important role in the analysis of the torque and power developed by a machine both in the steady-state and transient conditions.

The vector difference between the e.m.f. $\dot{E}_0$ due to the excitation flux and the terminal voltage $\dot{U}$ of a synchronous machine depends on the effect of the armature reaction and on the voltage drop in the active resistance and leakage inductive reactance of the armature winding.

Since armature reaction depends to a very great extent on the type of the machine (salient-pole or non-salient-pole), kind of load (inductive, active or capacitive) and on the degree of load symmetry (balanced or unbalanced), all these factors must be duly considered when plotting a voltage diagram.

It is necessary to bear in mind that all the e.m.f.s and voltages that participate as components in the voltage diagram should correspond to its fundamental frequency; therefore, all the e.m.f.s and voltages must preliminarily be resolved into harmonics and from each of them the fundamental wave must be taken separately 【1】. In the chapter where the armature reaction is considered, an analysis was carried out which allowed to obtain the fundamental voltage wave produced by the armature field components revolving in step with the machine rotor.

When a new machine is being commissioned, a vector diagram is plotted from the test data obtained from the experimental no-load and short-circuit characteristics.

The voltage across the terminals is the result of the action of the following factors: (a) the fundamental pole m.m.f. creating flux Φ_0 which induces the fundamental e.m.f. E_0; (b) the direct-axis armature reaction m.m.f. F_{ad} proportional to the load-current component I_d, reactive in respect to the e.m.f. E_0; (c) the quadrature-axis armature reaction m.m.f. F_{aq} proportional to the current I_q component, active in respect to the e.m.f. ; (d) the leakage e.m.f. $E_{\sigma a}=x_{\sigma a}I$, proportional to the load current I; (e) the active voltage drop in stator winding Ir_a. Since when $I=I_n$ the voltage drop Ir_a is less than 1% of the rated voltage, in most cases it may be neglected.

The diagram may be constructed by two different methods. In the first method of construction of the vector diagram it is assumed that each m.m.f. exists separately and produces its own magnetic flux, the latter creating its own e.m.f. Thus, four separate fluxes and, accordingly, four e.m.f.s created by them appear in the machine, viz: (a) the excitation flux $\dot{\Phi}_0$ and the fundamental e.m.f. $\dot{E}_0$; (b) the flux and the e.m.f. of the direct-axis armature-reaction $\dot{\Phi}_{ad}$ and $\dot{E}_{ad}$; (c) the flux and e.m.f. of the quadrature-axis armature-reaction $\dot{\Phi}_{aq}$ and $\dot{E}_{aq}$, and (d) the flux $\dot{\Phi}_{\sigma a}$ and the e.m.f. $\dot{E}_{\sigma a}$ of armature-winding leakage. If we also take into account the active voltage drop, which, when taken with opposite sign, may be considered as an e.m.f. $\dot{E}_r = -\dot{I}r_a$, the vector sum of the e.m.f.s listed above gives as a result the magnitude and phase of the terminal voltage vector $\dot{U}$.

Since the vector summation of fluxes and the corresponding e.m.f.s induced by them by the superposition method is legitimate only when the reluctances are constant in all sections of the magnetic circuit of the machine, this method is directly applicable to the unsaturated magnetic circuit of a synchronous machine. When using this method for machines with a saturated circuit, it is necessary to take into account the actual reluctances of the parts of the magnetic circuit at the given operating conditions and assume that the reluctances are constant so far as the given operating conditions are concerned [2]. The results obtained will be correct, nevertheless it is difficult to determine the real magnetic conditions of the machine.

Since by this method a vector summation of synchronous machine e.m.f.s is carried out, the voltage vector diagram obtained in this case may be called the e.m.f. diagram.

From the theoretical point of view, this diagram is of very great methodical importance, since it allows to access with the necessary completeness, the entire combination of factors determining, in the end, the voltage across the synchronous generator terminals, not with standing the fact that, for purposes of calculation and test, the diagram is somewhat complicated [3]. Therefore, for a series of practical purposes the e.m.f. diagram is given a number of modifications to bring it into a more simple and convenient form.

The method that is of greatest interest is the Blondel two-reaction theory, according to which all fluxes due to the load current $\dot{I}$, including leakage flux $\dot{\Phi}_{\sigma a}$, are resolved along the direct and quadrature axes. In connection with this, we introduce the concept of synchronous machine direct and quadrature-axes reactances x_d and x_q, and their components, which represent some of the fundamental parameters of a synchronous machine and serve for assessment of its performance characteristics.

By the second method we may determine first the resultant m.m.f. of the generator, obtained as the result of interaction of the excitation m.m.f. with a armature-reaction m.m.f., and, having found from it the resultant flux in the air gap $\dot{\Phi}_\delta$, then determine the e.m.f. $\dot{E}_\delta$ actually induced in the machine [4]. By subtracting vectorially from e.m.f. $\dot{E}_\delta$ the inductive voltage drop in the leakage reactance $j\dot{I}x_{\sigma a}$ and the active voltage drop $\dot{I}r_a$, we may find the resultant voltage across the generator terminals.

The diagram of m.m.f.s and e.m.f.s obtained in this case is called the Potier regulation, or e.m.f. diagram.

For balanced load conditions, assuming that the parameters of all phases are equal, we may restrict the construction to a diagram for one phase only.

It should be noted that the vector diagrams constructed for a synchronous generator operating as a generator may be readily extended to its operation as a synchronous motor and a synchronous condenser. The most simple voltage diagram is obtained for balanced load of a synchronous non-salient-pole generator with an unsaturated magnetic system. We shall therefore begin the discussion with the latter generator.

Let us construct the e.m.f. diagram of a synchronous non-salient-pole generator first for the case of inductive load, when $0<\varphi<90°$. Align the vector of the generator voltage across the generator terminals with the positive direction of the ordinate axis and draw the current vector $\dot{I}$ lagging behind the voltage vector $\dot{U}$ by an angle φ. Then draw the vector of the e.m.f. $\dot{E}_0$ produced by the magnetic excitation flux $\dot{\Phi}_0$, as leading the current vector I by an angle φ. According to the general rule, flux $\dot{\Phi}_0$ leads the e.m.f. vector $\dot{E}_0$ by $90°$.

The fundamental wave of the armature-reaction m.m.f. $\dot{F}_a$ of a synchronous generator rotates in step with its rotor. In a non-salient-pole type machine the difference between permeances along the direct axis and quadrature axes mat be neglected, and if may be assumed that m.m.f. $\dot{F}_a$ creates only a sine wave of reaction flux $\dot{\Phi}_a$. This flux coincides in phase with the current $\dot{I}$ and induces in the stator winding an e.m.f. $\dot{E}_a$ lagging in phase behind $\dot{I}$ by $90°$. If x_a is the inductive reactance of armature reaction for a non-salient-pole machine, then

$$\dot{E}_a = -j\dot{I}x_a$$

By vector addition of the flux vectors $\dot{\Phi}_0$ and $\dot{\Phi}_a$ and, respectively, the e.m.f. vectors $\dot{E}_0$ and $\dot{E}_a$ we obtain: (1) the vector of the resultant flux $\dot{\Phi}_\delta$ which actually exists in the generator air gap and determines the saturation of its magnetic circuit, and (2) the vector of the resultant e.m.f. $\dot{E}_\delta$ in the stator winding, proportional to flux $\dot{\Phi}_\delta$ and lagging behind it by $90°$.

Existing together with the armature-reaction flux is a stator winding leakage flux $\dot{\Phi}_{\delta a}$, the vector of which, like that of flux $\dot{\Phi}_a$, coincides in phase with current $\dot{I}$ and creates in the stator winding a leakage e.m.f. of fundamental frequency $\dot{E}_{\sigma a} = -j\dot{I}x_{\sigma a}$ a lagging in phase behind current I by $90°$. Here $x_{\sigma a}$ is the leakage reactance of the stator winding. Besides, it is necessary to take into account the e.m.f. $\dot{E}_r = -\dot{I}r_a$ which is opposite in phase to current $\dot{I}$; here r_a is the active resistance of the stator winding.

By vector addition of the e.m.f. vector $\dot{E}_0$, $\dot{E}_a$, $\dot{E}_{\sigma a}$ and $\dot{E}_r$, or, what is the same, the e.m.f.s $\dot{E}_\delta$, $\dot{E}_{\sigma a}$ and $\dot{E}_r$, we obtain the vector $\dot{U}$ of the voltage across the generator terminals. The angle φ by which current $\dot{I}$ lags behind voltage $\dot{U}$ is determined by the parameters of the external power circuit to which the generator is connected and feeds into. The vector $\dot{U}_c$ of the line voltage is in opposition to generator voltage vector $\dot{U}$.

When constructing the vector diagrams of a synchronous machine, it is not e.m.f.s $\dot{E}_a$, $\dot{E}_{\sigma a}$ and $\dot{E}_r$ that are represented, but their inverse values, which are the reactive and active voltage drops in the given sections of the circuit, i.e.,

$$-\dot{E}_a = j\dot{I}x_a, \quad -\dot{E}_{\sigma a} = j\dot{I}x_{\sigma a}, \quad -\dot{E}_r = \dot{I}r_a$$

In this case, the voltage diagram gives, obviously, the resolution of the e.m.f. $\dot{E}_0$ due to the excitation flux into components representing the voltage drops $j\dot{I}x_a$, $j\dot{I}x_{\sigma a}$ and $j\dot{I}r_a$ and the generator terminal voltage $\dot{U}$. On the other hand, the voltage diagram shows not the fluxes $\dot{\Phi}_0$, $\dot{\Phi}_a$ and $\dot{\Phi}_\delta$, but the e.m.f.s $\dot{F}_0$, $\dot{F}_a$ and $\dot{F}_\delta$, which produce them; this makes it legitimate to call it the e.m.f. diagram.

The voltage drop vectors $j\dot{I}x_a$ and $j\dot{I}x_{\sigma a}$ a may be substituted for by a common voltage drop vector

$$j\dot{I}x_a + j\dot{I}x_{\sigma a} = j\dot{I}x_s$$

where, the reactance $x_s = x_a + x_{\sigma a}$ is termed the synchronous reactance of the salient-pole machine. It is interesting to represent the relative arrangement in space of the main parts of a machine, the stator and the rotor, and the windings on them, together with the m.m.f.s they create. Angle φ indicates the space displacement of the conductors carrying maximum current $\dot{I}$ relative to the conductors which have the maximum e.m.f. $\dot{E}_0$ and are opposite the polepposite the poleame angle φ current $\dot{I}$ lags behind e.m.f. $\dot{E}_0$ in time phase 【5】. If we add the m.m.f. vector $\dot{F}_a$ to the excitation winding m.m.f. vector $\dot{F}_0$, we shall obtain the resultant m.m.f. vector $\dot{F}_\delta$ which lags in space from $\dot{F}_0$ by the same angle θ, by which the e.m.f. $\dot{E}_\delta$ lags behind e.m.f. $\dot{E}_0$ in time phase 【6】.

2.4.2 Specialized English Words

    balanced load　对称负载
    specifications　技术条件
    voltage across the terminals　端电压
    steady-state conditions　稳态
    in step with　与……同步
    active in respect to　相对……呈电阻性
    synchronous reactance　同步电抗
    percent variation of synchronous generator voltage　同步发电机的电压变化率
    initial machine design technical　电机的原始设计
    transient conditions　瞬态，暂态
    reactive in respect to　相对呈感性
    superposition method　叠加法
    in opposition to　方向或相位相反

synchronous condenser　同步调相机
fundamental performance characteristics　主要运行特性

2.4.3　Notes

【1】It is necessary to bear in mind that all the e.m.f.s and voltages that participate as components in the voltage diagram should correspond to its fundamental frequency; therefore, all the e.m.f.s and voltages must preliminarily be resolved into harmonics and from each of them the fundamental wave must be taken separately.

首先，应该看到 It 是形式主语，真正的主语是动词不定式短语 to bear in mind；而第一个 that 引导的是宾语从句，做 bear 的宾语；在该从句中，all the e.m.f.s and voltages 是主语，有 that 引导的 that participate as components in the voltage diagram 是该主语的定语从句；此句中的";"前后是两个独立的句子，它们之间既不是主从关系，也不是并列关系，therefore 是副词，在前后两句之间语义的逻辑关系中起推理的语气作用；therefore 后面的句子是一个由 and 连接的并列复合句。

【2】When using this method for machines with a saturated circuit, it is necessary to take into account the actual reluctances of the parts of the magnetic circuit at the given operating conditions and assume that the reluctances are constant so far as the given operating conditions are concerned.

首先，应该看到 When using this method for machines with a saturated circuit 是带疑问词 when 的分词短语做主句的状语，it 是主句的形式主语，真正的主语是两个并列的动词不定式 to take into...和（to）assume that；在后一个动词不定式中，由 that 引导的宾语从句 that the reluctances are constant so far as the given operating conditions are concerned，做动词不定式（to）assume that 的宾语；只是该从句本身又是一个主从复合句，其中由（so far）as 引导的 so far as the given operating conditions are concerned 是条件状语从句；so（as）far as sth. be concerned 即"就某事而言"。

【3】From the theoretical point of view, this diagram is of very great methodical importance, since it allows to access with the necessary completeness, the entire combination of factors determining, in the end, the voltage across the synchronous generator terminals, notwithstanding the fact that, for purposes of calculation and test, the diagram is somewhat complicated.

首先，句中有几个科技专业英语中常用的短语：From the theoretical point of view，即"从理论观点看"；is of very great methodical importance 即"具有重大的研究方法上的价值"；其次，从整句看，是一个由连词 since 引导的主从复合句，其中 notwithstanding the fact that, for purposes of calculation and test, the diagram is somewhat complicated 是介词 notwithstanding 引出的介词短语，在整句中做让步状语，只是该短语中介词的宾语 the fact 有一个同位语从句：that, for purposes of calculation and test, the diagram is somewhat complicated 对 the fact 进一步说明，说明是什么样的 fact；since 后面是原因状语从句，其句干为 it allows to access...the entire combination of factors...，至于分词短语 determining, in the end, the voltage across the synchronous generator terminals，则做 the entire combination of factors 的后置定语，其中的 in the end 是介词短语以插入语的形式在短语中做状语。

【4】By the second method we may determine first the resultant m.m.f. of the generator, obtained as the result of interaction of the excitation m.m.f. with a armature-reaction m.m.f., and, having found from it the resultant flux in the air gap $\varPhi_\delta$, then determine the e.m.f. E_δ actually induced in the machine.

这句尽管较长，但它是简单句，其句干是 we may determine first the resultant m.m.f...and... then determine the e.m.f. E_δ...。其中，obtained as the result of interaction of the excitation m.m.f. with a armature-reaction m.m.f 是过去分词短语做 the resultant m.m.f 的后置定语，词组 as the result of interaction of A with B 意为"A 与 B 相互作用的结果"; having found from it the resultant flux in the air gap $\varPhi_\delta$ 是现在分词（完成时）定语短语做后一个动词 determine 的时间状语，交代分词短语中的动词 found 的发生时间在谓语动词 determine 之前，是先从 it(the resultant m.m.f.) 求出 the resultant flux in the air gap $\varPhi_\delta$ 后，才 determine（确定）the e.m.f. E_δ。

【5】By this same angle φ current $\dot{I}$ lags behind e.m.f. $\dot{E}_0$ in time phase.

该句译为"电流 $\dot{I}$ 在时间相位上落后电势也是这个 φ 角"，正是为了强调同样是这个 φ 角，因此把 By this same angle φ 放在句首，实际上这句话的正常语序是 current $\dot{I}$ lags behind e.m.f. $\dot{E}_0$ by this same angle φ in time phase，其中介词 by 表示"到某种程度"。

【6】If we add the m.m.f. vector $\dot{F}_a$ to the excitation winding m.m.f. vector $\dot{F}_0$, we shall obtain the resultant m.m.f. vector $\dot{F}_\delta$ which lags in space from $\dot{F}_0$ by the same angle $\dot{\theta}$, by which the e.m.f. $\dot{E}_\delta$ lags behind e.m.f. $\dot{E}_0$ in time phase.

这个句子的结构还是比较清楚的，是一个主从复合句，从句是由 If 引导的条件状语从句；只是主句中的宾语 the resultant m.m.f. vector $\dot{F}_\delta$ 有一个定语从句 which lags in space from $\dot{F}_0$ by the same angle $\dot{\theta}$ 修饰它，而这个定语从句中又有一个定语从句 by which the e.m.f. $\dot{E}_\delta$ lags behind e.m.f. $\dot{E}_0$ in time phase. 修饰 the same angle $\dot{\theta}$。

2.4.4 Translation

三相同步发电机的运行分析

电压相量图对分析同步发电机的运行状况有着非常重要的作用。从电压相量图上，可以得到同步发电机的电压变化率，即由于负载减小电压的升高和发电机从空载到负载过程中电压的下降等。这些问题的解决具有重要意义：（1）在最初设计时要确定在不同的运行条件下所需的励磁电流；（2）在对已安装好的电机进行测试时，判定其性能是否与原技术设计书中一致。有了电压相量图，也使我们不必真实地施加负载，就能确定电机的运行状态。当电机的容量较大时，要真实地施加负载是特别困难的。

运用电压相量图，可通过计算得到电机的主要运行特性。最后，用电压相量图可以确定由励磁磁场产生的电势与端电压之间的功率角 θ。θ 无论在稳态还是在瞬态情况下，在分析电机产生的转矩和功率时都起着重要作用。

由励磁磁通产生的电势 $\dot{E}_0$ 与同步电机端电压 $\dot{V}$ 之间的相量差，取决于电枢反应的影响和电枢绕组的电阻及漏感抗上的电压降。

由于电枢反应的影响与电机的类型（凸极机还是隐极机），负载的性质（感性、电阻性还是电容性）和负载的平衡度（对称的还是不对称的）有很大关系，因此在画电压相量图时，所有这些因素都必须充分考虑到。

需要牢记的是，电压相量图中所有的电势和电压相量都是基波频率，因此所有的电势和电压都必须事先进行谐波分析，然后从这些谐波中将基波分离出来。在讨论电枢反应的那一章中，已经分析了可以获得由与电机转子同步旋转的电枢磁场分量产生的基波电压。

当一台电机将要交付使用时，技术人员根据从空载和短路特性实验中获得的数据绘出其电压相量图。

端电压是以下各因素共同作用的结果：(a) 基波磁极磁势产生磁通，进而感应基波电势 E_0；(b) 正比于负载电流的纵轴分量 I_d（其相对于电势 E_0 呈纯感性）的纵轴电枢反应磁势 F_{ad}；(c) 正比于负载电流的横轴分量 I_q（其相对于电势 E_0 呈纯电阻性）的横轴电枢反应磁势 E_{aq}；(d) 正比于负载电流 I 的漏电势 $E_{\sigma a} = x_{\sigma a} I$；(e) 定子绕组的纯电阻性电压降 Ir_a；由于当 $I=I_n$ 时，电压降 Ir_a 小于额定电压的 1%，因此在大多数情况下可以将其忽略。

电压相量图可由两种不同的方法绘制。在第一种方法中，假定各磁势相互独立存在，分别产生各自的磁通，它们的磁通又分别感应各自的电势。因此在电机中出现了由它们各自产生的 4 个磁通和 4 个电势，即：(a) 励磁磁通 $\dot{\Phi}_0$ 及其基波电势 $\dot{E}_0$；(b) 纵轴电枢反应磁通 $\dot{\Phi}_{ad}$ 及其电势 $\dot{E}_{ad}$；(c) 横轴电枢反应磁通 $\dot{\Phi}_{aq}$ 及其电势 $\dot{E}_{aq}$；(d) 电枢绕组漏磁通 $\dot{\Phi}_{\sigma a}$ 及其电势 $\dot{E}_{ad}$；如果再计及纯电阻压降，该压降当加上负号时，便可看作一电势 $\dot{E}_r = -\dot{I}r_a$，以上所有这些电势的相量和便确定了端电压相量 $\dot{U}$ 的大小和相位。

由于将磁通以及由它们感应的相应的电势，采用叠加法进行相量相加，只有当电机磁路的各横截面上的磁阻为常数时方能得到正确结果，因此这种方法直接适用于同步电机不饱和的磁路中。当将此方法适用于磁路饱和的电机时，必须将在给定运行条件下的磁路的各部分实际磁阻考虑进去，并就该给定的运行条件而言假设其磁阻为常数。尽管要确定电机的实际磁路状况是困难的，但这样做的结果将是正确的。

由于用此方法进行的是同步电机感应电势相量相加，因此在这种情况下得到的电压相量图便可称为电势相量图。

由于这种方法可以相当完整地测算各电势分量的相量和，最终确定同步电机的端电压，因此，从理论观点看，这种相量图具有重大的研究方法上的价值，尽管在用于计算机和实验目的时，该相量图是颇为复杂的。因此，对于许多实际用途，为使其更为简单和使用便利，人们对电势相量图进行了不少修正。

最有价值的方法是勃朗德的双反应理论，根据这个理论，由负载电流 $\dot{I}$ 产生的所有磁通，包括漏磁通 $\dot{\Phi}_{\sigma a}$，都沿纵轴和横轴分解。与此相关，我们引入同步电机纵轴和横轴电抗 x_d 和 x_q，以及它们的分量的概念，这些物理量代表同步电机某些基本参数，并且用于测算电机的运行特性。

用第二种方法，可先确定由励磁磁势与电枢反应磁势相互作用而得到的发电机的合成磁势，再由该合成磁势求出气隙磁通 $\dot{\Phi}_\delta$ 后，确定在电机中实际感应的电势 $\dot{E}_\delta$。通过从电势相量 $\dot{E}_\delta$ 中减去漏电抗的感应电势 $j\dot{I}x_{\sigma a}$ 和纯电阻压降 $\dot{I}r_a$ 后，便可求出发电机两端的合成电压。

在这种情况下得到的磁势和电势相量图称为保梯定则，或电势磁势相量图。

对于假定其各相参数均相等的对称负载情况，我们可以只画其中一相的相量图。

值得注意的是，作为发电机运行的同步发电机而画的相量图可以很容易地扩展到作为电动机和同步调相机运行时的情况。

最简单的相量图是磁路不饱和的隐极同步发电机带对称负载时所得到的。因此我们从这种发电机开始讨论。

首先画出感性负载（$0<\varphi<90°$）下隐极同步发电机的电势相量图。将发电机端电压相量$\dot{U}$置于坐标纵轴的正方向上，并在滞后其φ角处画电流相量$\dot{I}$。然后将由励磁磁通中$\dot{\Phi}_0$产生的电势相量$\dot{E}_0$画得领先电流相量$\dot{I}\varphi$角。根据定则，磁通$\dot{\Phi}_0$领先电势相量$\dot{E}_0$一个$90°$角。

同步发电机电枢反应磁势的基波$\dot{F}_a$与转子同步旋转。在隐极同步电机中，沿纵轴和横轴之间的磁导之差可以忽略不计，且可以假定磁势$\dot{F}_a$仅产生电枢反应正弦波磁通$\dot{\Phi}_a$。该磁通与电流$\dot{I}$同相，在定子绕组中产生滞后电流$\dot{I}$一个$90°$角的感应电势$\dot{E}_a$。若x_a为隐极电机电枢反应电抗，那么

$$\dot{E}_a = -j\dot{I}x_a$$

分别将磁通相量$\dot{\Phi}_0$与$\dot{\Phi}_a$、电势相量$\dot{E}_0$与$\dot{E}_a$相加，便得：(1) 合成磁通相量$\dot{\Phi}_\sigma$，该磁通是实实在在存在于发电机的气隙当中的，并决定磁路的饱和程度；(2) 定子绕组的合成电势相量$\dot{E}_\delta$，它正比于磁通$\dot{\Phi}_\delta$，并滞后其$90°$。

与电枢反应磁通同时存在的是定子绕组漏磁通$\dot{\Phi}_{\sigma a}$，与磁通相量$\dot{\Phi}_a$一样，该相量与电流$\dot{I}$同相，并在定子绕组中产生具有基波频率、在相位上滞后电流$\dot{I}$一个$90°$角的漏电势$\dot{E}_{\sigma a}=-j\dot{I}x_{\sigma a}$。式中$x_{\sigma a}$为定子绕组漏电抗，除此之外，还必须考虑电势$\dot{E}_r=-\dot{I}r_a$，该相量与电流$\dot{I}$在相位上相反，式中$r_a$为定子绕组纯电阻。

将电势相量$\dot{E}_0$、$\dot{E}_a$、$\dot{E}_{\sigma a}$和$\dot{E}_r$，或将电势相量$\dot{E}_\delta$、$\dot{E}_{\sigma a}$和$\dot{E}_r$进行相量相加，都得到发电机端电压相量$\dot{U}$。电流$\dot{I}$滞后于电压$\dot{U}$的φ角由与发电机相连并对其供电的外电路的参数决定。线电压相量$\dot{U}_c$与发电机电压相量$\dot{U}$的相位相反。

在画同步发电机的相量图时，图中的各相量并不是用电势相量$\dot{E}_a$，$\dot{E}_{\sigma a}$和$\dot{E}_r$来表示的，而是用它们的相反的量，也就是在给定电路中的感性和电阻性电压降来表示，即

$$-\dot{E}_a = j\dot{I}x_a, \quad -\dot{E}_{\sigma a} = j\dot{I}x_{\sigma a}, \quad -\dot{E}_r = \dot{I}r_a$$

显然，在这种情况下的电压相量图中，由励磁磁通产生的电势$\dot{E}_0$分解成代表电压降$j\dot{I}x_a$，$j\dot{I}x_{\sigma a}$，$j\dot{I}r_a$以及发电机端电压$\dot{U}$。另一方面，电压相量图并不显示磁通$\dot{\Phi}_0$，$\dot{\Phi}_a$和$\dot{\Phi}_\sigma$，而是显示产生这些磁通的磁势$\dot{F}_0$，$\dot{F}_a$和$\dot{F}_\delta$，这便使得称其为电势磁势相量图合情合理。

电压降相量$j\dot{I}x_a$和$j\dot{I}x_{\sigma a}$可由一公共的电压降相量来替代

$$j\dot{I}x_a + j\dot{I}x_{\sigma a} = j\dot{I}x_s$$

式中，电抗

$$x_s = x_a + x_{\sigma a}$$

称为凸极电机的同步电抗。将电机的各主要部分，即定转子及其绕组、与这些绕组产生的磁势在空间的相对位置表示出来是很有意思的。φ角表示电流$\dot{I}$达最大值的那相导体与电势$\dot{E}_0$取得最大值，且与磁极轴线相反的那相导体之间的空间位置差角。电流$\dot{I}$在时间相位上滞后于电势$\dot{E}_0$。同样是这个φ角，若将磁势相量$\dot{F}_a$与励磁绕组磁势相量$\dot{F}_0$相加，便可得到合成相量$\dot{F}_\delta$，该磁势在空间相位上滞后$\dot{F}_0$的角正是电势$\dot{E}_\sigma$在时间相位上滞后电势$\dot{E}_0$的同一个相位角θ。

Unit 2.5 Motor Control Based on Microcomputer

2.5.1 Text

　　Computer automation of factories, homes, and offices is ushering a new era of industrial revolution. This automation of the future will significantly advance our industrial civilization and profoundly influence the quality of human life on this planet. Microcomputer-based intelligent motion control systems which constitute the workhorses in the automated environment will play a significant role in the forthcoming era.

　　Electronic motion control technology has moved a long way since the introduction of power semiconductor devices in the mid-1950s. In course of its dynamic evolution during the last three decades, the area of motion control has grown as diverse interdisciplinary technology. The frontier of this technology has taken a new dimension with the advent of today powerful microcomputers, VLSI circuits, power integrated circuits, and advanced computer-added design (CAD) techniques.

　　The paper gives a comprehensive review of state-of-the-art motion control technology in which the salient technical features of electrical machines, power electronic circuits, microcomputer control, VLSI circuits, machine controls and computer-aided design techniques have been discussed, and wherever possible, appropriate trends of the technology have been indicated.

　　Microcomputer-based intelligent motion control systems are playing a vital role in today industrial automation. In an automated industrial environment, a hierarchical computer system makes decisions about actions based on a present strategy, and a motion control system, as a workhorse, translates these decisions into mechanical action.

　　Today motion control is an area of technology that embraces many diverse disciplines, such as electrical machines, power semiconductor devices, converter circuits, dedicated hardware signal electronics, control theory, and microcomputers, more recently, the advent of VLSI/ULSI circuits and sophisticated computer-aided design techniques has added new dimensions to the technology 【1】. Each of the component disciplines is undergoing an evolutionary process, and is contributing to the total advancement of motion control technology. The motion control engineer today is indeed facing a challenge to keep abreast with this complex and ever-growing multidisciplinary technology.

　　Motion control is a term defined by the present generation of engineers. It is an offspring of electrical machine drives technology, which has grown at a rapid pace over the last two decades. The era of electronic motion control essentially started with the advent of power semiconductor devices in the late 1950s, though hydraulic, pneumatic and other mechanically driven actuation systems were known for a long time. Gradually, the use of integrated signal electronics simplified the electronic control hardware. The introduction of microcomputers in the early 1970s profoundly influenced motion control systems, not only by simplifying the control hardware, but by adding intelligence as well as diagnostic capability to the system.

　　We have seen an explosive growth in the application of motion control systems during recent years. Mechanical motion control systems found widespread acceptance in industry since the invention of the steam engine started the first industrial revolution in eighteenth century, when mass industrial manufacturing replaced manual labor 【2】. Since then, the evolution of motion control engineering has been influenced by the development of electrical machines, vacuum tube electronics, gas tube electronics, saturable reactor magnetics solid-state electronics and control theory. The advent of computer technology and microelectronics during recent years has brought us to the doorstep of a second industrial revolution. Today, a tremendous momentum has developed for computer automation of our factories, homes, and offices. The principal motivation for this automation is improvement of productivity and quality and minimization of less predictable human elements; and these motives in turn are being inspired by international competition.Computer-aided design (CAD) and computer-aided manufacturing (CAM) are playing increasingly important roles in factory automation. The concept of computer-integrated manufacturing(CIM), in which business decisions are translated to designs, which are then translated to manufacturing through a hierarchy of computers and motion control systems, will become a reality in the near future 【3】.

A motion control system, as mentioned before, is the workhorse through which higher level computer decisions are translated into mechanical actions. Motion control applications in industry include robots, numerically controlled machine tools, general-purpose industrial drives, computer peripherals, and instrument type drives, in the home applications include home appliance drives for washers, dryers, air-conditioners, blenders, mixers, etc. In a typical computer-controlled manufacturing system on a factory floor, there are three layers of control. The master control (usually a minicomputer) operates the entire network. It includes parts transportation and material handling on machine tools by robots. The direct numerical control (DNC) unit, usually a second minicomputer, collects programs for the microcomputers which directly control the machine tools. The computerized numerical control units (CNC), in addition, contain diagnostic programs that can detect mechanical and electronic malfunctions in a machine tool and report them to central controllers. The data entry units allow communication between the operator and the DNC computer.

In motion control systems, the application of robots is of significant interest today. The robot essentially symbolizes the challenge of synthesizing all state-of-the art component technologies. The modern industrial robot was introduced by Japan in 1980, and since then, it has evolved from performing simple tasks, such as handling and transferring, to performing sophisticated work including welding, painting, assembling, inspection, and adjustment. In Japan, the world leader in factory automation, almost two hundred thousand robots are in operation today. This is about 60 percent of all industrial robots in the world. One noticeable trend is the growth of robot use in non-manufacturing fields, for example, nuclear power generation, medical service, welfare, agriculture, construction, transport and warehousing, underwater work, and space exploration. More intelligent robots that will mimic the brain and muscles of human beings will be put to work in the future, for factory, home, or office automation.

The application of motion control has growth at a phenomenal rate in the computer peripheral industry. For example, in the U.S.alone, electronic printers, disk drives and tape drives used 24 million motors in 1983, and this figure is expected to rise to a staggering 80 million by the year

1988.It has been estimated that an average American home uses 50 motors in all the household appliances, and this amounts to a staggering 12.5 billion motors in all U.S.homes. Eventually, all these motors will be controlled by microcomputer. In an automated home of the future, all the motors will have a central home computer-based control through an integrated power-and-signal wiring system. Similar integrated motion control concepts will be applied to automobiles, airplanes, and so on.

This report is intended to review the technology trends of motion control that relate to electrical machines, power semiconductor devices, converter circuits, microcomputers, VLSI circuits, control of machines, and computer-aided control design techniques. Particular emphasis will be paid to intelligent motion control based on microcomputers. Again, motion control systems that use call machines will be our main there of discussion.The literature on motion control has grown enormously, and proliferated so diversely that it is impossible to deal with all the aspects of the technology. Therefore, only the salient features will be highlighted.

2.5.2 Specialized English Words

computer automation of factory, home and office 工业、家庭和办公的计算机自动化
industrial civilization 工业文明
microcomputer-based 建立在微机控制基础上
automated environment 自动化领域
ULSI（Ultra Large Scale Integration） 特大规模集成电路
invention of the steam engine 蒸汽机的发明
vacuum tube electronics 真空管电子学
gas tube electronics 气体管电子学
control theory 控制理论
Computer-integrated Manufacturing（CIM） 计算机集成制造
general-purpose industrial drive 通用工业驱动装置
instrument type drive 仪表类的驱动装置
home appliance drive 家电驱动装置
direct numerical control（DNC） 直接数字控制
nuclear power generation 核能发电
disk drive 磁盘驱动器
integrated motion control 集成运动控制
a new era of industrial revolution 工业革命的新纪元
quality of human life 人类的生活质量
intelligent motion control system 智能运动控制系统
Computer-added Design（CAD） 计算机辅助设计
VLSI（Very Large Scale Intergration） 超大规模集成电路
electrical machine drive technology 电机驱动技术
integrated signal electronics 集成信号电子学

first industrial revolution　第一次工业革命
saturable reactor magnetics　饱和电抗器电磁学
solid-state electronics　固体电子学
Computer-aided Manufacturing（CAM）　计算机辅助制造
robot　机器人
numerically controlled machine tool　数控机床
computer peripheral　计算机外设
washer，dryer，air-conditioners blender，mixer　洗衣机、甩干机、空调、搅拌器、混合器
computerized numerical control（CNC）　计算机化的数字控制
electronic printer　电子打印机
tape drive　磁带驱动器

2.5.3　Notes

【1】Today motion control is an area of technology that embraces many diverse disciplines, such as electrical machines, power semiconductor devices, converter circuits, dedicated hardware signal electronics, control theory, and microcomputers, more recently, the advent of VLSI/ULSI circuits and sophisticated computer-aided design techniques has added new dimensions to the technology.

该句子比较长，也较为复杂，因此有必要看清楚该句的主框架结构：Today motion control is an area of technology..., the advent of...has added new dimensions to...，其意为"当今世界的运动控制是一个……技术领域，……的出现拓宽了……的新的运用范围"。那么当今世界的运动控制是一个什么样的技术领域呢？这就是由关系词 that 引导的定语从句 that embraces many diverse disciplines, such as electrical machines, power semiconductor devices, converter circuits, dedicated hardware signal electronics, control theory, and microcomputers 所说明的意思"其包括诸如电机、电力半导体装置、转换器电路、专门的硬件信号电子学、控制理论、微型计算机等方方面面的众多技术领域"的这么一个综合性的技术领域。那么又是什么样的东西的出现拓宽了谁的新的运用范围呢？more recently, the advent of VLSI/ULSI circuits and sophisticated computer-aided design techniques,（has added new dimensions）to the technology，"最近出现的超大规模集成电路，甚至于特大规模集成电路，再加上日渐成熟的计算机辅助设计技术出现"，"又进一步拓宽了运动控制技术领域新的运用范围"。由此可见，该句应翻译为："当今世界的运动控制是这么一个综合性的技术领域，其包括诸如电机、电力半导体装置、转换器电路、专门的硬件信号电子学、控制理论、微型计算机等方方面面的众多技术领域。最近出现的超大规模集成电路，甚至于特大规模集成电路，再加上日渐成熟的计算机辅助设计技术出现，又进一步拓宽了运动控制技术领域新的运用范围。"

【2】Mechanical motion control systems found widespread acceptance in industry since the invention of the steam engine started the first industrial revolution in eighteenth century, when mass industrial manufacturing replaced manual labor.

翻译该句时主要是要注意句中的 when mass industrial manufacturing replaced manual labor 是非限定性定语从句，而不是时间状语从句，其意为"当时大规模的工业生产化代替了手工

劳动"。其作用是进一步说明 the invention of the steam engine 的,即说明"蒸汽机发明"的那个时代,当时大规模的工业生产化代替了手工劳动。而 the first industrial revolution 则是由 since 引导的时间状语从句的主谓结构 the invention of the steam engine started 的宾语,in eighteenth century 则是该时间状语从句中的时间状语。故全句应该翻译为:"18世纪蒸汽机的发明开启了第一次工业革命,当时大规模的工业生产化代替了手工劳动。从此以后,机械的运动控制系统便在工业领域得到广泛应用"。

【3】The concept of computer-integrated manufacturing(CIM), in which business decisions are translated to designs, which are then translated to manufacturing through a hierarchy of computers and motion control systems, will become a reality in the near future.

该句的结构较为复杂,其主句框架是 The concept...will become a reality in the near future,意为"……的概念将在不远的将来成为现实"。什么样的概念呢?of computer-integrated manufacturing(CIM),"计算机集成制造的概念"。计算机集成制造的概念又是什么呢?in which business decisions are translated to designs,"在此概念中,首先是将商务决策转化为产品设计"。然后又怎么样呢?which are then translated to manufacturing through a hierarchy of computers and motion control systems,"然后又通过计算机和运动控制系统将其转化为产品制造"。因此该句应该译为:"计算机集成制造的概念,将在不远的将来成为现实。所谓计算机集成制造,就是首先将商务决策转化为产品设计,然后又通过计算机和运动控制系统将其转化为产品制造"。

2.5.4 Translation

基于微型计算机的电动机控制

工业、家庭和办公系统的计算机自动化,开创了工业革命的新纪元。在不远的将来,这种自动化的实现将大大加快工业化发展的步伐,深刻地影响着生活在这个星球上的人类的生活质量。建立在微机控制基础上的智能运动控制系统,将成为自动化领域中的主力军,并将在下一个世纪中起着举足轻重的作用。

基于电子技术的运动控制技术,自20世纪50年代中期电力半导体装置的采用至今,已走过了漫长的发展之路。运动控制在过去30年的动力性能的改进过程中,不断地延伸到了各种相关技术领域中。随着当今功能强劲的微机技术、超大规模集成电路、电力集成电路,以及计算机辅助设计等技术的迅速发展,运动控制这一领域的运用范围已大大扩展。

本文就运动控制技术进行了全面的阐述,其中对电机、电力电子电路、微机控制、超大规模集成电路、机械控制,以及计算机辅助设计技术等的卓越性能进行了讨论,并尽可能地对该技术的大致发展趋势作出了评述。

以微机为基础的智能运动控制系统,在当今工业自动化领域中起着至关重要的作用。在自动化的工业环境中,根据当前运行情况下的应对策略,多级微机系统对下一步的动作作出决策,而运动控制系统作为其执行机构,则将该决策转化为机械运动。

当今世界的运动控制是这么一个综合性的技术领域,其包括诸如电机、电力半导体装置、转换器电路、专门的硬件信号电子学、控制理论、微型计算机等方方面面的众多技术领域。最近出现的超大规模集成电路,甚至于特大规模集成电路,再加上日渐成熟的计算机辅助设计技术出现，又进一步拓宽了运动控制技术领域新的运用范围。上述的每一技术领域均在不断地发展，这又对运动控制技术的进步作出贡献。当今世界的运动控制领域内的工程师们，的确面临着赶上这个复杂的、至今还在不断发展、多学科的综合性技术领域的前进步伐的严峻挑战。

"运行控制"是当今一代的工程师们命名的，其为电机驱动技术的产物，在近20年间得到了迅速发展。尽管此前很长一个时期内，人们熟知的是以液压、气动或者其他的机械式驱动方式，但从20世纪50年代后期电力半导体装置的出现起，电子式运动控制方式便基本上开始起步了。集成信号电子学的应用逐渐地简化了电子控制的硬件。20世纪70年代早期微型计算机的应用，由于其不仅简化了系统的控制硬件，而且在系统中引入智能控制技术和故障诊断能力，因而对运行控制系统产生了根本性的影响。

近年来运动控制系统应用已经得到了迅猛的发展。18世纪蒸汽机的发明开启了第一次工业革命，当时大规模的工业生产化代替了手工劳动。从此以后，运动机械的运动控制系统便在工业领域得到广泛应用。再由此往后，运动控制系统的发展便一直受到电机学、真空管电子学、气体管电子学、饱和电抗器电磁学、固体电子学以及控制理论发展的影响。近年来计算机技术和微电子学的发展，开启了人类历史上的第二次工业革命的里程碑。当前，工业、家庭、办公的计算机自动化正以惊人之势迅猛发展。发展自动化的主要目的是为了提高生产率和产品质量，是为了最大限度地节省有限的人力资源，而在这方面的国际竞争又反过来大大地刺激了这一目的的实现。计算机辅助设计和计算机辅助制造在工业自动化中占着越来越重要的地位。计算机集成制造的概念，将在不远的将来成为现实。所谓计算机集成制造，就是首先将商务决策转化为产品设计，然后又通过计算机和运动控制系统将其转化为产品制造。

如上所述，运动控制系统是将上一级的计算机决策，转化为机械运动的一种控制机构。运动控制在工业上的应用包括机器人、数控机床、通用工业驱动装置、计算机外设和仪表类的驱动装置，在家庭的应用包括运用于洗衣机、甩干机、空调、搅拌器、混合器等家用电器中的家电驱动装置。通常出现在工厂车间里的计算机控制制造系统，均具有三层控制。主控制（一般为微机）控制整个系统，包括通过机器人完成零部件的传送，以及对机床上的加工材料的处置和掌控。直接数字控制单元（一般为第二级微机），则汇集直接控制机床的微机的控制程序。此外，在计算机化的数字控制单元中，还包含可发现机床的机械故障和电子故障的诊断程序，一旦发现将向中央控制器发出报告。数据输入单元可实现操作人员与直接数字控制单元的计算机进行通信。

当前，在运动控制系统中，机器人的应用具有极其重要的价值。可以说，机器人基本上对所有的人工智能技术都可谓是巨大的挑战。现代工业机器人最初于1980年在日本诞生，从那以后，其由执行抓握和传输物件这样一些简单任务，不断改进为完成诸如焊接、喷漆、装配、检验和调节等复杂的工作。在日本这样一个工业自动化水平处于世界领先地位的国度里，现有20万台机器人投入了运行，占全世界工业机器人总量的60%。机器人在诸如核能发电、医疗服务、社会福利、农业生产、基础建设、交通运输、仓库装卸、水下作业、空间探测等非制造领域中的应用推广，已是一个很明显的发展趋势。不久的将来，能模仿人类大脑思维和肌肉运动功能的更具智能的机器人，将在工业、家用和办公自动化中投入使用。

在计算机外围设备中，运动控制系统的应用正以非凡的速度迅猛地发展。例如，1984年仅仅在美国一个国家，电子打印机、磁盘驱动器和磁带驱动器所使用的电机就有2千4百万台，这个数字预计到 1988 年，将剧增至 8 千万。据估计，平均每个美国家庭的家用电器设备中总共要用到 50 台电机，也就是说，所有美国家庭所运用的电机数总计为 125 亿台。到头来，所有这些电机又终将由微机来控制。在将来的现代化家庭中，所有电机都通过一集成的有线供电和传输信号的方式，由计算机控制中心进行控制。类似的集成运动控制概念，还将在汽车和飞机等这样一些交通工具上得到应用。

　　本文旨在对与电机、电力半导体装置、转换器电路、微型计算机、超大规模集成电路、机械控制，以及计算机辅助设计等有关的技术发展方向进行评述，其侧重点是以微型计算机为基础的智能运动控制。此外，用于指挥机械运动的控制系统也是其讨论的重点。有关运动控制方面的文章种类繁多，且又在不断地更新，因而不可能一一介绍，在此仅介绍其主要特点。

Part 3   Electric Generation, Distribution & Power System

Unit 3.1   Major Electrical Plant in Power Station

Unit 3.2   Operation & Control of Power System

Unit 3.3   Relaying Protection in Power System

Unit 3.4   SCADA & EMS in Power System

Unit 3.5   Electric Power Flow Calculation in Power System

Unit 3.1 Major Electrical Plant in Power Station

3.1.1 Text

Introduction

In the last nine chapters we have studied the different types of electrical power stations from the various basic points of view, arriving at principles for the choice of plant, its main dimensions and layout. The main dimensions of generators of different types were also considered. However, the details of the major electrical equipment in the various types of power stations, their characteristics, the choice of their detailed specifications, etc., must also be studied. In this chapter, the main electrical plant required in power stations is discussed and overall specifications are given for guidance in the choice of the required plant.

The major electrical plant in power stations consists of generators, exciters for the generators, power transformers, reactor, circuit-breakers, switch-board and control-board equipment. In addition, station transformers, auxiliary equipment, emergency lighting equipment, etc., are required. Details of the protective equipment-relays will be discussed in a separate chapter.

Generators and Exciters

The main of generators and exciters have already been considered. Some of the more detailed specifications of generators which should be considered in making a final choice are discussed in this section.

Generators Constants

In addition to the normal rating and other specified quantities, the constants of the generators should be known to enable one to determine their performance in the power system under different loading conditions at different power factors, under transient conditions and under fault conditions; they are also used in studies of system stability 【1】. In addition to the normal continuous rating, specifications given so far have included kVA rating, number of phases, frequency, voltage, power factor, star connection of the stator and temperature rise limits.

In the analysis of salient-pole machines in particular, as well as in that of synchronous machines in general, the two reaction theory is used. The direct-axis synchronous reactance X_d and quadrature-axis X_q, give the performance of a synchronous machine under normal and symmetrical conditions. The direct-axis synchronous reactance is the ratio of the fundamental component of reactive armature voltage, due to the fundamental direct-axis component of armature current, to this component of current under steady-state conditions and at rated frequency 【2】. In the same way, the quadrature-axis synchronous reactance is the ratio of the fundamental component of reactive armature voltage, due to the fundamental quadrature-axis component of armature current, to this component of current under steady-state conditions and at rated frequency. For cylindrical rotor machines, such as turbo alternators, X_d and X_q are identical. For salient-pole machines X_q is about 0.6 to 0.7 times X_d. The synchronous reactance, X_d and X_q, are useful in determining the regulation of a synchronous generator and also the short-circuit currents under symmetrical short-circuit conditions.

The short-circuit ratio is another important factor. This is the ratio of the field current at rated open circuit voltage and rated frequency to the field current at rated armature current on sustained symmetrical short-circuit at rated frequency. For large generators short-circuit ratio varies between 0.8 and 1.2.

When studying the performance of a synchronous machine under sudden short-circuit conditions, its behavior under transient conditions should be determined. With the sudden symmetrical short circuit of an alternator on no-load, the armature current decreases during the first few cycles. With a prolonged short-circuit, the current finally attains the steady value $I_d = E/X_d$, where E is the synchronous internal voltage and X_d is the synchronous reactance. If the envelope of the current wave is projected back to the zero time—the instant of starting the short-circuit, then, neglecting the first few cycles during which the decrement is very rapid, the value of current obtained is the transient current, $I' = E'/X'_d$, where E' is the transient internal voltage and X'_d is the transient reactance. The sub-transient current is the value of the current at zero time, given by $I'' = E''/X''_d$, where E'' is the sub-transient internal voltage and X''_d is the sub-transient reactance. The increment of armature current represented by the sub-transient component is due to induced current in the damper wingding. The transient reactance of a synchronous machine is equal to the armature leakage reactance.

Machine	X_d	X'_d	X''_d	X_2	X_0
Steam turbo-generator	95 to 145	12 to 27	7 to 18	7 to 18	1 to 15
Water-wheel generator(no damper winding)	60 to 145	20 to 47	17 to 40	30 to 79	4 to 24
Salient-pole generator(with damper winding)	60 to 145	20 to 52	13 to 36	13 to 36	2 to 20
Synchronous machine used as a compensator	150 to 225	27 to 55	18 to 38	18 to 38	2 to 316

The time-constants of a synchronous machine give an idea of the time of decay in the voltage or current components which help in setting the relays for protection under particular conditions of faults, etc.Some of the time constants useful in the studies are mentioned below with their definitions.

The direct-axis transient open-circuit time constant, T'_{d0}, is the time in seconds required for the root mean square (r.m.s) a.c. value of the slowly decreasing component present in the direct-axis component of symmetrical armature voltage on open circuit to decrease to 1/e or 0.368, of its initial value when the field wingding is suddenly short-circuited with the machine running at rated speed [3].

The direct-axis transient short-circuited time constant, T'_d, is the time in seconds required for the r.m.s.a.c value of the slowly decreasing component present in the direct-axis symmetrical a.c.

component of armature current under suddenly applied 3-phase short-circuit conditions, with the machine running at rated speed, to decrease to 1/e or 0.368 of its initial value.

The direct-axis sub-transient short circuit time constants, T_d'', is the time in seconds required for the rapidly decreasing component present during the first few cycles in the direct axis a.c. component of the armature current under suddenly applied 3-phase short-circuit conditions, with the machine running at rated speed, to decrease to 1/e or 0.368, of its initial value. The short circuit time constant of the armature winding, T_d, is the time in seconds for the asymmetrical component of armature current under suddenly applied short-circuit conditions, with the machine running at rated speed, to decrease to 1/e or 0.368, of its initial value.

When studying the performance of a synchronous machine under asymmetrical fault conditions such as a line-to-line or line-to-earth fault, the negative and zero sequence reactance of the machine should be known to enable one to work out the fault current and voltage conditions of the machine. The negative sequence reactance, X_2, and the zero sequence reactance, X_0, are, therefore, other important constants of the synchronous machine which should be given in the detailed specifications. The negative sequence reactance, X_2, is the ratio of the fundamental component of reactive armature voltage, due to the fundamental negative-sequence component of armature current, to this component of armature current at rated frequency. The zero sequence reactance, X_0, is the ratio of the fundamental component of reactive armature voltage, due to the fundamental zero-sequence component of armature current, to this component of armature current at rated frequency.

Power Transformer Type and Characteristics

The power transformers used in power station and power systems are either single-phase banks of three each or 3-phase transformers. When single-phase transformers are operating in parallel, their voltage rating should be identical, their percentage impedance voltages should be equal and their ratios of reactance to resistance must be equal. This also applies to a bank of three single-phase transformers. Departure from the above ideal conditions of parallel working results in uneconomical load which the bank can carry. It is not a good practice to operate transformers in parallel under the following conditions.

(1) When the load division is such that the total load is equal to the combined kVA rating, the load current flowing in one of the transformers exceeds 110% of the normal full-load current.

(2) When the no-load circulating current, exclusive of excitation current, exceeds 10% of the full-load current.

(3) When the arithmetic sum of the circulating current and the load currents exceeds 110% of the normal full-load current.

Transformer Connections

Three-phase transformers can be used with advantage in place of banks of three single-phase transformers. The types of connection for the 3 phase operation of transformers normally considered are delta-star, star-delta and delta-delta. Star-star connections are not normally used. The merits and demerits of different types of connection are discussed below. When compared different types of transformer connections for 3-phase operation, the degree of voltage symmetry, the voltage and current harmonics and any peculiarities of the connections should be considered. When comparing three single-phase units with a 3-phase unit, the ratio of kVA-output/kVA-rating of the bank should be considered.

Balanced 3-phase delta-delta, star-delta and delta- star banks do not introduce third harmonics or their multiples into the line. The wave shapes of the magnetizing currents are superior compared to those of symmetrical banks【4】. With the delta connection, there is current and voltage symmetry 【5】. A system occurs only with open-delta and T-connections when only two single-phase transformers are used in a 3-phase system to supply part of a 3-phase load【6】. The star connection gives symmetrical values as far as line currents are connected but introduces third-harmonic current and voltage between line and neutral, and has the possibility of developing dangerous over-voltages.

Generally, the delta connection is used for transformer windings when the voltage is small and the currents to be handled are large. This is because the phase voltage is the same as the line voltage and phase current is equal to the line current divided by $\sqrt{3}$. When the voltages are large and the current is small, i.e. for high-voltage windings, it is preferable to use star connection. Here the phase voltage is equal to the line voltage divided by $\sqrt{3}$ and the phase current is the same as the line current. Thus, with the star connection, the windings can be insulated for lower voltage. The delta-star connection is used if the neutral is to be taken out for loading or for earthing, the star connection being on the secondary side. The question whether to use delta-delta or delta-star connections in a particular power system is decided by the need for parallel operation with existing banks of transformers or the interconnection of the network or system apparatus.

When comparing delta-delta and delta-star banks of transformers, some of the following points are to be noted. The voltage of delta-delta transformers are completely determined by the circuit characteristics. The division of current depends on the internal characteristics of transformers. The division of load is equal only if the individual transformers have equal impedances. In delta-star the division of current is entirely independent of the difference in the characteristics of the individual transformers. The balanced 3-phase load is equally divided among the phases in delta-star banks regardless of inequalities in impedance.

The effect of difference in voltage ratios is marked in the case of delta-delta bank of transformers. Large currents flow in both high-voltage and low-voltage windings if all phases are not alike. By the proper proportioning of impedances, the division of load can be made to be in the same proportion as the ratings of the transformers. Delta-star transformers are insensitive to a difference in voltage ratios.

A delta-delta bank of transformer is versatile. The failure of one single-phase transformer out of the three leaves two transformers connected in open delta, or vee, so that a 3-phase supply can still be obtained. The capacity of the bank is, however, reduced to 86.6% of that of the two working units, or 57.7% of that of the three. With two transformer units connected in open delta initially supplying a certain load, it is possible to add a third unit and connect the bank delta to supply more load as it grows in the system. With delta-delta transformers, the failure of one phase is more serious and may involve discontinuity of supply.

With star-star connection there is instability of the neutral, for a number of reasons. The instability is due to differences between the magnetizing in different phases which occurs owing to differences in the magnetic circuits by faults in construction 【7】. Also, the neutral is shifted owing to the unbalanced load from line to neutral. The potential of the physical neutral is generally at some point other than the geometric center of the voltage triangle, and is greatly affected by the characteristics of the load and the other circuit conditions. In star-connected transformers, third harmonic magnetizing currents cannot flow.

With star-delta or delta-star transformers, the presence of the delta eliminates the third-harmonic voltages associated with the star connection and stabilizes the neutral. The presence of the star connection makes the division of load current among the phases independent of impedance. Thus units having different impedances can be used in a star-delta bank. However, the kVA ratings of the individual phases are unequal and the maximum safe output of the bank is three times the rated kVA of the smallest transformer.

Vector Groups in Transformer Connection

The polarity of a transformer denotes the phase displacement which exists between its primary and secondary phase voltage. If the windings for the primary and secondary are wound in the same direction and sense from the terminals ends, there is no phase displacement between the voltages and the transformer polarity is termed as subtractive polarity. If the windings are wound in opposite directions, there is 180° phase displacement between the voltages and the transformer polarity is additive polarity.

For satisfactory parallel operation of single phase transformers, the polarity of the transformers must be considered for proper connections.

In case of three phase transformers, the phase displacement between the primary and secondary voltages may occur due to the type of connections. The main types of transformer connections were discussed in the above section. If the clock hands represent the displacement between primary and secondary voltage vectors, the phase difference is shown from 0 to 360 by the position of hands e.g. Ⅰ shows -30°, Ⅺ shows +30°; 0 shows no phase displacement; 6 shows 180° phase displacement, etc.

The vector groups in common use for three phase power transformers are given by I.S: 2026—2962 and for distribution transformers by IS: 1180—1964. There are mainly the following groups:

Groups No.1	Phase displacement between h.v and l.v. winding voltage=0°
Yy0	Star/ Star
Dd0	Delta/Delta
Dz0	Delta/Zigzag
Groups No.2	Phase displacement=180
Yy6	Star/ Star
Dd6	Delta/Delta

续表

Groups No.1	Phase displacement between h.v and l.v. winding voltage=0°
Dz6	Delta/Zigzag
Groups No.3	Phase displacement = minus 30
Dy1	Star/Star
Yd1	Delta/Delta
Yz1	Delta/Zigzag
Groups No.4	Phase displacement = plus 30
Dy11	Star/Star
Yd11	Delta/Delta
Yz11	Delta/Zigzag

The relative merits and demerits of different types of connections for power transformers were discussed in the above section. For distribution transformers, the most common type of vector group used in Dy11 Delta/Star.

For satisfactory parallel operation of three phase transformers, the vector group of transformers must be the same, or they would match in such a way that there would be no phase displacement between the voltages on the h.v. and l.v. side. Connection of h.v. is indicated by capital letter, connection of l.v. is indicated by small letter and the figures 0, 1, 6, 11 indicate phase displacement as described by hands of a clock.

3.1.2 Specialized English Words

electrical plant 电气设备
reactor 电抗器
switch-board 开关屏
exciter 励磁机
temperature rise limit 温升极限
direct-axis 直轴
sudden short-circuit condition 突然短路
transient internal voltage 次暂态内电势
damper wingding 阻尼绕组
time of decay 衰减时间
direct-axis transient open-circuit time constant 直轴暂态开路时间常数
direct-axis sub-transient short circuit time constant 直轴次暂态短路时间常数
negative and zero sequence reactance 负序和零序电抗
circulating current 环流
3-phase transformer 三相芯式变压器
third harmonics or their multiples 三及三的倍数次谐波
excitation current 励磁电流

subtractive polarity　减极性
vector groups in transformer connection　变压器的连接组别
power station　发电厂
circuit-breakers　断路器
control-board　控制屏
station transformer　厂用变压器
salient-pole machine　凸极电机
quadrature-axis　交轴
transient condition　暂态
sub-transient　次暂态
sub-transient reactance　次暂态同步电抗
armature leakage reactance　电枢漏电抗
direct-axis transient short-circuited time constant　直轴暂态短路时间常数
asymmetrical fault　不对称故障
line-to-line fault　相对相故障
line-to-earth fault　相对地故障
single-phase banks of three each　单相组式（变压器）
transient over-voltage　暂态过电压
percentage impedance voltage　阻抗电压百分数
symmetrical banks　呈放射状的（磁路），即三相芯式变压器
voltage ratio　电压比
additive polarity　加极性

3.1.3　Notes

【1】In addition to the normal rating and other specified quantities, the constants of the generators should be known to enable one to determine their performance in the power system under different loading conditions at different power factors, under transient conditions and under fault conditions; they are also used in studies of system stability.

该句为一并列复合句，只不过连接两子句的不是并列连词，而是标点符号";"。后一个子句比较简单，只要清楚了 they 是指代 the constants 就应该没有什么问题了。只是前一个子句就比较复杂了，主要是出现了一连串的由介词短语组成的条件状语 under different loading conditions、at different power factors、under transient conditions 和 under fault conditions，显得其修饰关系比较复杂。但在翻译英语中的同一种状语（比如此处均为条件状语）时，只要记住一点，那就是始终是从后翻译到前。究其原因，就是原英语中的条件状语出现的前后顺序是由小（小条件）到大（大条件），而汉语则与之相反，是由大（大条件）到小（小条件）。因此该句应译为："除了通常所说的额定值和其他一些特定的物理量以外，还应该了解发电机的参数，以便使我们能够确定，无论电力系统是在暂态情况下，还是在故障情况下，其处于不同的功率因数下的不同负荷条件下的性能。这些参数也同样运用于系统的稳定性研究"。

【2】The direct-axis synchronous reactance is the ratio of the fundamental component of reactive armature voltage, due to the fundamental direct-axis component of armature current, to this component of current under steady-state conditions and at rated frequency.

该句的主框架是 The direct-axis synchronous reactance is the ratio of the fundamental component...to this component of current...。其意为"直轴同步电抗是……的基波分量与在……（条件下的）电流的该分量的比值"。什么样的基波分量呢？of reactive armature voltage, due to the fundamental direct -axis component of armature current，即"由电枢反应电流的直轴基波分量所产生的电枢反应基波电压"。在什么条件下的该分量呢？under steady-state conditions and at rated frequency，即"在额定频率和稳态条件下"。因此该句译为"直轴同步电抗，是由电枢反应电流的直轴基波分量所产生的电枢反应基波电压，与在额定频率和稳态条件下基波电流分量的比值"。

【3】The direct-axis transient open-circuit time constant, T'_{d0}, is the time in seconds required for the root mean square (r.m.s) a.c. value of the slowly decreasing component present in the direct-axis component of symmetrical armature voltage on open circuit to decrease to 1/e or 0.368, of its initial value when the field wingding is suddenly short-circuited with the machine running at rated speed.

该句比较长，结构也比较复杂，不容易翻译好。首先应该看清楚，该句的句干是 The direct-axis transient open-circuit time constant, T'_{d0}, is the time...，其意思很清楚，就是"直轴暂态开路时间常数 T'_{d0} 是一个……（什么样的）时间"。那么究竟是什么样的时间呢？那就是用两个连续的短语来说明内容，其分别为介词短语 in seconds 和过去分词短语 required for the root mean square (r.m.s) a.c. value of the slowly decreasing component present in the direct-axis component of symmetrical armature voltage on open circuit to decrease to 1/e or 0.368, of its initial value when the field wingding is suddenly short-circuited with the machine running at rated speed。别看第二个短语很长，其中还含有从句，但就语法而言其只不过是一个短语。这也就是说，这是"一个以秒计的，为……所需要的时间"。那么是为谁所需要的呢？for the root mean square (r.m.s) a.c. value，"交流电的均方根值（所需要的）"。该值是谁的呢？of the slowly decreasing component，"一个缓慢衰减的分量"。该分量出现在何处呢？present in the direct-axis component of symmetrical armature voltage on open circuit，"出现在开路时的对称电枢电压的直轴分量上"。这个缓慢衰减的分量衰减到何值呢？to decrease to 1/e or 0.368, of its initial value，衰减到"为其初始值的 1/e，也就是 0.368"之时。那么该初始值又是一个什么样的值呢？这就是由定语从句 when the field wingding is suddenly short-circuited with the machine running at rated speed 引出的定语"当电机以额定转速运行，励磁绕组突然短路时（的值）"。因此将以上分析连贯起来，该句便可译为："直轴暂态开路时间常数 T'_{d0} 是一以秒计的时间，其为出现在开路时三相对称电枢电压的直轴分量上的、那个缓慢衰减的交流电的均方根值，衰减到为其初始值的 1/e，也就是初始值的 0.368 之时所需要的时间"。

【4】The wave shapes of the magnetizing currents are superior compared to those of symmetrical banks.

该句原本并不难翻译，只是句中 symmetrical banks 究竟是何意？就得思量一番。symmetrical 的原意是"呈放射状的"，banks 的原意是"几个分支"。此处是在讲述三相变压器的磁路，故而可知 symmetrical banks 就是讲"呈放射状的几个分支磁路"，这不就是三相芯式变压器吗？可见 symmetrical banks 应该译成"三相芯式变压器"。因此全句应该译为："单相组式变压器的激磁电流的波形要好于芯式变压器"。

【5】With the delta connection, there is current and voltage symmetry.

该句看似简单，但其却蕴涵着很多与专业知识有关的信息。首先 With 就不能简单地译为"具有"，或者"如果具有"，而应该结合专业知识，译为"三相变压器的原、副边只要有一边是"。因此全句应译为："三相变压器只要有一边是三角形接法，其电流和电压就都是对称的"。

【6】A system occurs only with open-delta and T-connections when only two single-phase transformers are used in a 3-phase system to supply part of a 3-phase load.

首先应该看出，该句的句干为 A system occurs only...when...，其意很清楚"某系统只会出现在……的时候"。而 with open-delta and T-connections 是一介词短语做"A system"的后置定语，只不过其二者之间隔了一个谓语动词 occurs。这是英语中的一种修饰语与被修饰语分离的现象，在英汉翻译时要引起注意。究其原因，就是因为该后置定语，相比谓语动词 occurs（仅有一个单词）太长，因此让谓语动词先出现，以免头重脚轻。此外，就是一个时间状语 when only two single-phase transformers are used in a 3-phase system to supply part of a 3-phase load 的翻译问题了，应该不难。所以全句应译为："在三相系统中，只有在用两台单相变压器来承担部分三相负荷时，三相变压器中才会出现开口的三角形和 T 形的接法"。

【7】The instability is due to differences between the magnetizing in different phases which occurs owing to differences in the magnetic circuits by faults in construction.

该句的主框架为 The instability is due to differences between，其意很清楚，即"这种不稳定性是由于……之间的差别而引起的"。究竟是什么东西之间的差别呢？the magnetizing in different phases，"各相之间的激磁作用"的差别。而这种激磁作用的差别是怎么产生的呢？owing to differences in the magnetic circuits by faults in construction，"由于结构上的误差所造成的各相之间的磁路不对称"而产生的。故全句应译为："这种不稳定性，是由于各相之间的激磁作用的差别而引起的，而这种差别又是由于结构上的误差所造成的各相之间的磁路不对称所产生的"。

3.1.4　Translation

电站的主要电气设备

引言

在前几章中，已从各种基本观点研究了不同类型的发电厂，得出了设备选择的基本原理，即设备的主要容量和配置的选择。各种不同类型的发电机的容量也给予了考虑。然而，各种不同类型的发电厂中主要电气设备的具体细节，即这些设备的特性，以及在选择这些设备时应具备哪些要求也必须加以研究。本章将讨论发电厂中所需的主要电气设备，其中所有具体的结论，将作为选择所需设备的指导性意见。

发电厂中的主要电气设备包括：发电机、发电机的励磁机、电力变压器、电抗器、断路器、开关屏和控制屏等设备。此外，还配备了厂用变压器、辅助设备、紧急照明设备等。保护设备也就是继电器将在专门的章节中具体讨论。

发电机和励磁机

发电机和励磁机的主要问题此前已经讨论过了。本节将要讨论的问题是在最终决定选择发电机时应考虑的一些更具体的细节。

发电机的参数

除了通常所说的额定值和其他一些特定的物理量以外，还应该了解发电机的参数，以便我们能够确定，无论电力系统是在暂态情况下，还是在故障情况下，其处于不同的功率因数下的不同负荷条件下的性能。这些参数也同样运用于系统的稳定性研究。除了通常所说的那些连续运行时的额定值外，目前还需特别说明的技术数据包括额定容量（千伏安）、相数、频率、电压、功率因数、定子绕组的星形连接，以及温升极限等。

在具体分析凸极电机，以及进行同步电机的一般性分析时，都要用到双反应理论。同步电机的直轴同步电抗 X_d 和交轴同步电抗 X_q，表征了在正常和同步条件下同步电机的运行特性。同步电机的直轴同步电抗，是由电枢反应电流的直轴基波分量所产生的电枢反应基波电压，与在额定频率和稳态条件下的基波电流分量的比值。同理，其交轴同步电抗，则是由电枢电流交轴基波分量所产生的电枢反应基波电压，与在额定频率和稳态条件下的基波电流分量的比值。对于隐极式转子的同步电机来说，比如汽轮交流发电机，其直轴同步电抗 X_d 与交轴同步电抗 X_q 是相等的。对于凸极式转子的同步电机来说，其 X_q 大约是 X_d 的 0.6～0.7 倍。同步电抗 X_d 和 X_q，在确定对称短路条件下同步发电机的调节特性，以及短路电流是非常有用的。

短路比是另一个非常重要的参数。其为在额定开路电压和额定频率下的励磁电流，与在额定频率下的稳态对称短路的条件下，电枢电流达到额定时的励磁电流的比值。对于大型发电机设备，短路比在 0.8～1.2 之间变化。

在研究同步发电机在突然短路条件下的特性时，应该了解同步发电机在暂态条件下的表现。当交流发电机在空载情况下突然发生对称的突然短路时，电枢电流在起始的几个周波内是减小的。随着短路状态的持续，电枢电流最终将趋于稳态值 $I_d = E/X_d$，式中，E 为同步电机的内电势，X_d 为同步发电机的直轴同步电抗。假如电流波形的包络线投影到零时刻，也就是起始短路的那一瞬间，那么，忽略波形下降变化很快的最初那几个周波，所得到的便是暂态电流，$I' = E'/X'_d$，式中，E' 为发电机的暂态内电势，X'_d 为暂态同步电抗。次暂态电流是在零时刻的电流值，其由式 $I'' = E''/X''_d$ 给出，式中，E'' 为发电机的次暂态内电势，X''_d 为次暂态同步电抗。由次暂态分量表现出来的电枢电流的增大部分，是由于在阻尼绕组中产生了感应电流才出现的。同步电机的暂态电抗等于电枢漏电抗。

电机类型	X_d	X'_d	X''_d	X_2	X_0
汽轮发电机	95～145	12～27	7～18	7～18	1～15
水轮发电机（无阻尼绕组）	60～145	20～47	17～40	30～79	4～24
凸极电机（带阻尼绕组）	60～145	20～52	13～36	13～36	2～20
用作调相机的同步电机	150～225	27～55	18～38	18～38	2～16

同步电机给出了一个时间常数，在出现某些特殊故障等条件下，有助于整定继电器的电压和电流分量的衰减时间的概念。一些在研究过程中很有用的时间常数及其定义简述如下。

直轴暂态开路时间常数 T'_{d0} 是一个以秒计的时间，其为出现在开路时三相对称的电枢电压直轴分量上的、那个缓慢衰减的交流电的均方根值，衰减到为其初始值的 1/e，也就是 0.368 之时所需要的时间。

直轴暂态短路时间常数 T'_d 是一以秒计的时间，其为出现在发生三相突然短路情况时的三相对称交流电枢电流的直轴分量上的、那个缓慢衰减的交流电的均方根值，在电机以额定转速运行条件下，衰减到为其初始值的 1/e，也就是 0.368 时所需的时间。

直轴次暂态短路时间常数 T''_d 是一个以秒计的时间，其为出现在发生三相突然短路情况时的交流电枢电流的直轴分量上的、那个迅速衰减的电流分量，在电机以额定转速运行条件下，在最初的几个周波内，衰减至初始值的 1/e，也就是 0.368 时所需的时间。电枢绕组的短路时间常数 T_d 是一个以秒计的时间，其为在电机以额定转速运行时突然发生三相短路的情况下，电枢电流的非对称分量衰减至初始值的 1/e，也就是 0.368 时所需的时间。

在研究同步电机在不对称故障条件下的特性，如相对相、相对地等故障情况时，应了解同步电机的负序和零序电抗，以便分析得出电机的故障电流和电压的情况。因此，负序电抗 X_2 和零序电抗 X_0 是另一些应该具体了解的同步电机的重要参数。负序电抗 X_2 是由电枢电流的基波负序分量所产生的电枢反应电压的基波分量，与在额定频率下的电枢电流的基波分量的比值。零序阻抗 X_0 是由电枢电流的基波零序分量所产生的电枢反应电压的基波分量，与在额定频率下的电枢电流的基波分量的比值。

电力变压器的类型和特性

在电站或电力系统中所使用的电力变压器，既可以是单相组式变压器，也可以是三相芯式变压器。当单相变压器并联运行时，其额定电压和阻抗电压百分数均应该相等，且其电抗与电阻的比率也必须相等。以上所述也同样适用于单相组式变压器。若违背上述变压器的并联运行理想条件，其结果是将会导致各变压器负荷分配不经济。在变压器并联运行时，以下情况均为不当之举：

（1）变压器并联运行时出现这种情况，即变压器组所承担的总负荷与其总的额定千伏安相等，但其中某一台变压器所流过的电流，却超出额定的满负荷电流的 110%。

（2）空载时的环流（除励磁电流外）超出满负荷电流的 10%。

（3）变压器内的环流与负荷电流的算术和，超出额定的满负荷电流的 110%。

变压器的连接方法

采用三相芯式变压器代替三相变压器组具有许多优点。变压器三相运行时，通常可给予考虑的连接方式为三角形/星形、星形/三角形和三角形/三角形。星形/星形连接方式通常不采用。变压器不同连接方式的优点和缺点将在下面加以讨论。在比较三相运行的变压器不同类型的接线方式时，应该将电压的对称度、电压和电流的谐波，以及接线方式的特点等加以考虑。在将三相组式变压器与三相芯式变压器进行比较时，应考虑组式变压器的输出千伏安数与其额定的千伏安数的比率。

对称的三相三角形/三角形、星形/三角形、三角形/星形这三种连接方式，都不会在输电线路上产生三次及三的倍数次谐波。单相组式变压器的激磁电流的波形要好于芯式变压器。三相变压器的原、副边只要有一边是三角形接法，其电流和电压就都是对称的。在三相系统中，只有在用两台单相变压器来承担部分三相负荷时，在三相变压器中才会出现开口的三角形和 T 形的接法。星形连接方式只要出现线电流就是对称的，但是在相线与中性线间便会出现三次谐波电流和电压,并有可能会演变成危险的过电压。

一般来说,当电压较低而所载电流又较大时,变压器的绕组通常采用三角形接法。这是因为相电压与线电压相等,而相电流则等于线电流除以$\sqrt{3}$。当电压较高而线电流又较小时,例如高压绕组,便优先采用星形接法。在此接法中,相电压等于线电压除以$\sqrt{3}$,而相电流等于线电流。这样,对于星形接法来说,便可采用低压的绝缘等级。假如不需要用中性点来接负载或接地,一般原边就采用三角形接法,副边就采用星形接法。在某一具体的电力系统中,究竟是采用三角形/三角形接法,还是采用三角形/星形接法的问题,要在现有的变压器组的具体情况,或是其与电力系统中的其他电力设备的连接情况下,根据并联运行的需要来决定。

在比较变压器的三角形/三角形与三角形/星形接法时,应该注意以下几个问题。三角形/三角形接法的变压器绕组,其电压完全由电路的特性所决定。而电流的分配则取决于变压器的内部特性。只有在每台变压器有相同的阻抗时,负荷的分配才会相同。在三角形/星形的接法中,电流的分配与每台变压器的特性差异完全无关。在三角形/星形接法的变压器组中,就是其阻抗不相等,其三相平衡负荷在各相中也是均衡分配的。

在三角形/三角形接法的变压器的情况中,电压比之间的差别所产生的影响比较明显。如果所有各相的电压比不相同的话,不管是在高压绕组还是在低压绕组中都会产生很大的电流。通过将变压器阻抗的比例关系调整到适当的值,负荷便会按与变压器额定值的相同比例来分配。对于不同的电压比,三角形/星形接法的变压器是不敏感的。

三角形/三角形变压器组的接法在运用中较易于变通。当该种接法的三相变压器中有一台单相变压器出现故障,剩下两台变压器便可接成开口三角形,也就是 V 形接法,此时仍然能提供三相供电。只是变压器组的容量减少到两台变压器的 86.6%,或三台变压器容量的 57.7%。当系统中负荷增长时,最初供应一定负荷的开口三角形变压器组(由两台变压器连接成),可以增加第三台变压器,从而接成三角形变压器组,便可供应更多的负荷。对于三角形/三角形接法的变压器,单相故障所造成的危害性会更大,并且可能造成停电事故。

由于种种原因,星形/星形接法的变压器组的中性点是不稳定的。这种不稳定性,是由于各相之间的激磁作用的差别而引起的,而这种差别又是由于结构上的误差,所造成的各相之间的磁路不对称所产生的。当然,若某相对于中线之间存在着不平衡负荷,所谓中性点也就荡然无存了。中性点的物理电势位一般是在电压三角形内的某一点之上,而不是在其几何中心上。且该点的位置受负荷的特性和其他电路条件的影响很大。星形接法的变压器,三次谐波电磁电流不能通过。

而对于星形/三角形或三角形/星形接线的变压器来说,由于三角形接法的出现,因而消除了星形接法时所具有三次谐波电压,也就稳定了中性点的位置。星形接法的出现,使各相之间的负荷电流的分配与阻抗无关。因此,各相的阻抗值不相同的变压器,可采用星形/三角形接法。然而,此时各相中额定千伏安容量是不相等的,变压器组的最大的安全输出容量为最小的变压器的额定容量(千伏安)的 3 倍。

变压器的连接组别

变压器的极性,表明了变压器原边和副边的相电压的相位差。若变压器的原边和副边绕组是按同一方向缠绕的,且都从末端口往里看,那么变压器的原、副边电压的相位差为零,而此极性称之为减极性。若变压器原、副边绕组的缠绕方向相反,那么变压器原、副边电压的相位差为 180°,而此极性称之为加极性。

为了能使单相变压器顺利地并联运行，对于变压器的具体连接方式下的极性必须认真加以考虑。在三相变压器的情况下，由于变压器原、副边的接线方式的不同，可能会出现原、副边电压之间的相位差。变压器接线的主要类型在上一节中讨论过了。若采用时钟的指针来表示变压器原、副边的电压相量的相位差，那么时钟的指针可以表示出 0°～360°的角度差，即用 1:00 表示-30°的相位差，11:00 钟表示+30°的相位差，0:00 表示无相位差，6:00 表示相差为 180°的相位差等。

三相电力变压器通常采用的连接组别，由国际标准 2026—2962 给出，配电变压器的则由国际标准 118—1964 给出。以下各组是主要的一些连接组别。

第一组：	高低压绕组间电压的相差为 0°
Yy0	星形/星形
Dd0	三角形/三角形
Dz0	星形/Z 形
第二组：	高低压绕组间电压的相差为 180°
Yy6	星形/星形
Dd6	三角形/三角形
Dz6	星形/Z 形
第一组：	高低压绕组间电压的相差为-30°
Dy1	星形/星形
Yd1	三角形/三角形
Yz1	星形/Z 形
第一组：	高低压绕组间电压的相差为+30°
Dy11	星形/星形
Yd11	三角形/三角形
Yz11	星形/Z 形

有关电力变压器不同连接方式的优缺点在上节中已经讨论过了。对于配电变压器，最常用的相量组别是使用三角形/星形 11:00 的连接组别。

为了三相变压器能顺利地并联运行，变压器的连接组别必须相同，或者说其高低压侧为这样一种接线的匹配方式，即在变压器的高低压侧电压之间没有相角差。高压侧的接线方式用大写字母表示，低压侧的接线方式用小写字母表示，数字 0、1、6、11 表示用时钟的指针所表示的相角差。

Unit 3.2　Operation & Control of Power System

3.2.1　Text

The purpose of a power system is to deliver the power the customers require in real time, on demand, within acceptable voltage and frequency limits, and in a reliable and economic manner【1】. In normal operation of a power system, the total power generation is balanced by the total load and transmission losses. The system frequency and voltage on all the buses are within the required limits, while no overloads on lines or equipment are resulted. However, loads are constantly changed in small or large extents, so some control actions must be applied to maintain the power system in the normal and economic operation state.

Optimal Economic Operation

It is an important problem how to operate a power system to supply all the (complex) loads at minimum cost. The basic task is to consider the cost of generating the power and to assign the allocation of generation to each generator to minimize the total "production cost" while satisfying the loads and the losses on the transmission lines【2】. The total cost of operation includes fuel, labor, and maintenance costs, but for simplicity the only variable costs usually considered are fuel costs. The fuel-cost curves for each generating unit are specified, the cost of fuel used per hour is defined as a function of the generator power output. With hydro-generation, however, in dry periods, the replenishment of the water supply may be problem. The water used today may not be available, in the future when its use might be more advantageous. Even without the element of the prediction involved, the problem of minimizing production cost over time becomes much more complicated.

It should be mentioned that economy of operation is not only possible consideration. If the "optimal" economic dispatch requires all the power to be imported from a neighboring utility through a single transmission link, considerations of system security might preclude that solution. When water used for hydro-generation is also used for irrigation, non-optimal releases of water may be required. Under adverse atmospheric conditions it may be necessary to limit generation at certain fossil-fuel plants to reduce emissions.

In general, costs, security and emission are all areas of concern in power plant operation, and in practice the system is operated to effect a compromise between the frequently conflicting requirements.

Power System Control

Power system control is very important issue to maintain the normal operation of a system. System voltage levels, frequency, tie-line flows, line currents, and equipment loading must be kept within limits determined to be safe in order to provide satisfactory service to the power system customers.

Voltage levels, line currents, and equipment loading may vary from location within a system, and control is on a relatively local basis. For example, generator voltage is determined by the field current of each particular generating unit; however, if the generator voltages are not coordinated, excess var flows result. Similarly, loading on individual generating units is determined by the throttle control on thermal units or the gate controls on hydro-units. Each machine will respond individually to the energy input to its prime mover. Transmission line loadings are affected by power input from generating units and their loadings, the connected loads, parallel paths for power to flow on other lines, and their relative impedances.

Active Power and Frequency Control

For satisfactory operation of a power system, the frequency should remain nearly constant. Relatively close control of frequency ensures constancy of speed of induction and synchronous motors. Constancy of speed of motor drives is particularly important for satisfactory performance of all the auxiliary drives associated with the fuel, the feed-water and the combustion air supply system. In a network, considerable drop in frequency could result in high magnetizing currents in induction motors and transformers. The extensive use of electric clocks and the use of frequency for other timing purpose require accurate maintenance of synchronous time which is proportional to integral. A change in active power demand at one point is reflected throughout the system by a change in frequency. Because there are many generators supplying power into system, some means must be provided to allocate change in demand to the generators. A speed governor on each generating unit provides the primary speed control function, while supplementary control origination at a central control center allocates generation.

In an interconnected system with two or more independently controlled areas, in addition to control of frequency, the generation within each area has to be controlled so as to maintain scheduled power interchange. The control of generation and frequency is commonly referred to as load-frequency control (LFC).

The control measures of power and frequency include:
(1) Regulation of the generator speed governor
(2) Underfrequency load shedding
(3) Automatic generation control (AGC)

AGC is an effective means for power and frequency control in large-scale power systems. In an interconnected power system, the primary objectives of AGC are to regulate frequency to the specified nominal value and to maintain the interchange power between control area at the scheduled values by adjust the output of the selected generators [3]. This function is commonly referred to as load-frequency control. A secondary objective is to distribute the required change in generation among units to minimize operating costs.

Reactive Power and Voltage Control

For efficient and reliable operation of power system, the control of voltage and reactive power should satisfy the following objectives:
(1) Voltages at the terminals of all equipment in the system are within acceptable limits. Both utility equipment and customer equipment are designed to operate at a certain voltage rating.

Prolonged operation of the equipment at voltages outside the allowable range could adversely affect their performance and possibly cause them damage.

(2) System stability is enhanced to maximize utilization of the transmission system.

(3) The reactive power flow is minimized so as to reduce the equipment and the transmission lines losses to a practical minimum. This ensures that the transmission system operates efficiently, i.e. mainly for active power transfer.

The problem of maintaining voltages within the required limits is complicated by the fact that the power system supplies power to a vast number of loads and is fed from many generating units. As loads vary, the reactive power requirements of the transmission system vary. Since reactive power cannot be transmitted over long distance, voltage control has to be effected by using special devices dispersed throughout the system. This is in contrast to the control of frequency which depends on the overall system active power balance. The proper selection and coordination of equipment for controlling reactive power and voltage are among the major challenges of power system engineering.

The control of voltage levels is accomplished by controlling the production, absorption, and flow of reactive power at all levels in the system. The generating units provide the basic means of voltage control; the automatic voltage regulators control field excitation to maintain a scheduled voltage level at the terminals of the generators. Additional means are usually required to control voltage throughout the system. The devices used for this purpose may be classified as follows:

(1) Sources of reactive power, such as series capacitors, shunt reactors, synchro-nous condensers, and static var compensators (SVCs).

(2) Line reactance compensators, such as series capacitors.

(3) Regulating transformers, such as tap-changing transformers and boosters.

Synchronous condensers and SVCs provide active compensation; the reactive power absorbed/supplied by them is automatically adjust so as to maintain voltages of the buses to which they are connected.

3.2.2 Specialized English Words

optimal economic operation 最优经济运行
fuel-cost curve 燃耗成本曲线
hydro-unit 水轮发电机组
load-frequency control (LFC) 负荷频率控制
automatic voltage regulator 电压自动调节器
series capacitor 串联电容器
synchro-nous condensers line reactance compensator 线路电抗补偿器
tap-changing transformer 分接头可调变压器
generating unit 发电机组
thermal unit 汽轮发电机组
throttle control （蒸汽）阀门控制

gate control　（水流）闸门控制
speed governor　（水轮机的）调速器
automatic generation control　（AGC）发电自动化控制
sources of reactive power　无功功率源
shunt reactor　并联电抗器
SVC（static var compensator）　静止无功补偿器
regulating transformer　可调压变压器
booster　升压器

3.2.3　Notes

【1】The purpose of a power system is to deliver the power the customers require in real time, on demand, within acceptable voltage and frequency limits, and in a reliable and economic manner.

该句并不长，也不甚复杂，但在翻译时还有几处要处理好。首先是整个句子的结构要看清楚，其主框架是 The purpose...is to deliver...in...and in...，其意为"目的是以……方式和……方式，向……提供……"。谁的目的呢？of a power system，"电力系统的"。向谁提供呢？the customers require，"用户的需求"。提供什么呢？the power，"电力"。以什么方式呢？in real time...and in a reliable and economic manner，"适时的和经济可靠的方式"。只不过所谓"适时"是什么意思呢？这就是 in real time 后面所插入的介词短语 on demand, within acceptable voltage and frequency limits 所说明的内容，其意为"一旦需要，就在其（用户）所能接受的电压和频率的范围内"。再加上适当的转译，本句应译为："电力系统运行的目的就是一旦用户需要，便能以适时的、可靠的、经济的方式，向用户按其所要求的电压和频率的范围提供其所需的电力"。

【2】The basic task is to consider the cost of generating the power and to assign the allocation of generation to each generator to minimize the total "production cost" while satisfying the loads and the losses on the transmission lines.

首先还是看该句的主框架：The basic task is to consider...and to assign...，"基本任务就是考虑……和分配……"。要考虑的是什么呢？这就是系动词后所接的两个动词不定式所组成的表语：to consider the cost of generating the power and to assign the allocation of generation to each generator，"考虑发电成本和将发电量分配到每台发电机"。怎样分配到每台发电机呢？to minimize the total production cost ...and the losses on the transmission lines，"使总的生产成本和每条传输线路上的损耗降到最小"。降低总的生产成本有没有前提呢？有，那就是 while satisfying the loads，"在满足负荷需求的同时"。因此本句应译为："要达到这个目的，其基本的任务就是考虑发电成本，以及在满足负荷需求的同时，将总发电量合理地分配到每台发电机，以使各台发电机总的生产成本和每条传输线路上的损耗降到最小"。

【3】In an interconnected power system, the primary objectives of AGC are to regulate frequency to the specified nominal value and to maintain the interchange power between control area at the scheduled values by adjust the output of the selected generators.

该句的修饰与被修饰之间的关系有一点复杂。首先还是看清楚句子的主框架：the primary objectives...are...，其意很清楚，就是"首要的目标是……"。什么地方的首要的目标呢？In an interconnected power system,"在相互连接的电力系统中"。谁的首要的目标呢？of AGC,"发电的自动化控制"。首要的目标是什么呢？to regulate frequency...and to maintain the interchange power...,"调节频率……和维持……的功率交换"。将频率调节到什么程度呢？to the specified nominal value,"到指定的额定值"。维持什么东西与什么东西之间的功率交换呢？between control area,"各控制区域之间的"。什么样的功率值呢？at the scheduled values,"（生产）计划中的功率值"。怎样进行调节呢？by adjust the output of the selected generators,"通过调节所选定的发电机的输出功率"来进行调节。综合以上分析，再做一些译文上的文字调整，可见该句应译为："在相互连接的电力系统中，发电自动化控制的首要的目标有两个，一是将频率调节到指定的额定值，二是通过调节所选定的发电机的输出功率，维持各控制区域之间的计划发电量之间的交换"。

3.2.4 Translation

电力系统的运行与控制

电力系统运行的目的就是一旦用户需要，便能以适时的、可靠的、经济的方式，向用户按其所要求的电压和频率的范围提供其所需的电力。在电力系统正常运行时，其所总发电量与总负荷及输电损耗之和是相平衡的。由于线路和设备没有出现过载，因而系统中所有母线的电压和频率均在所需的范围之内。然而，负荷总是经常在或大或小地变化，因而必须采取一些控制手段，以使电力系统维持在一个稳定和经济的运行状态。

最优经济运行

怎样才能使电力系统以最为经济的方式，为所有的复杂的负荷提供电能，是一个很重要的问题。要达到这个目的，其基本的任务就是考虑发电成本，以及在满足负荷需求的同时，将总发电量合理地分配到每台发电机，以使各台发电机总的生产成本和每条传输线路上的损耗降到最小。运行总成本包括燃料消耗、劳动力、维护费用，但为使问题得以简化，通常认为唯一会变化的成本是燃料消耗。每一台发电机组的燃耗成本曲线都是不同的，每小时的燃耗成本便定义为发电机功率输出的一个函数。然而对于水轮发电机来说，在干旱的季节水的补给可能是一个问题。今天用掉了水，今天就不再有水了，而明天也许就多得用不完。即使没有水文预报所涉及的问题，在整个时间周期内，生产成本最小化也是相当复杂的。

应该指出的是：经济运行并不是唯一可能的考虑，如果"最优"经济调度要求所有的能量都由一条单支路的传输线路，从临近的电力设备中传送过来，那么出于系统安全性的考虑，该方案可能会被否决。当运用于水力发电的水源还同时用于农业灌溉时，那么也许就需要放弃发电的最优经济运行方案。在空气质量不佳环境下，限制燃油或燃煤的发电厂的发电量，以减少燃烧尾气的排放也是必需的。

通常说来，成本、安全和排放问题是电厂运行需要考虑到的全部问题。而系统的实际运行，就是在这些常常是相互矛盾的需求之间，不断地进行协调的过程。

电力系统控制

电力系统控制是维持电力系统正常运行非常重要的问题。为了给电力系统的用户提供令人满意的服务，系统的电压、频率、连接线处的涌流、线电流，以及设备负荷率，必须要限制在安全的范围之内。

系统中的电压、线电流和设备负荷率在各处的情形可能是不相同的，会因所处的位置而异。因而，电力系统的控制也就应该因地制宜。比如说，发电机的电压由各自的发电机组的励磁电流所决定。然而，如果某台发电机电压因为励磁电流的问题而与系统的电压有所差异，便会在系统中另外产生无功潮流。同样，每台机组的负荷，在汽轮发电机组中是由蒸汽阀门控制决定的，在水轮发电机组中是由水流闸门控制决定的。每一台机组都要分别对来自于其原动机的能量输入作出反应。输电线的负荷情况，受到发电机组及其负载、连接的负荷、与之并联的其他线路上的涌流，以及其相关的阻抗的影响。

有功功率与频率控制

为了能使电力系统的运行得到令人满意的效果，频率应该维持大致恒定。频率的相关控制确保了电磁感应电动机和同步电动机转速的恒定。电动机转速的恒定对于那些与像燃料、给水和燃烧系统的送风等有关的辅机设备的良好运行来说是尤其重要的。在电网中，当频率下降到一定程度，就会引起感应电动机和变压器中励磁电流增大。电子钟和其他运用频率计时的装置的广泛应用，都要求维持准确的同步时间，而时间是与频率的积分成正比的。系统中某一处的有功功率需求的变化，是系统频率变化的反映。因为系统中有许多台发电机在向电网提供有功功率，所以必须将这些变化按照需求分别分配到各台发电机。每台发电机组的水轮机的调速器负责对原动机的速度调节，而中央控制中心则进行辅助性的控制，以给每台发电机组分配供电量。

在由两个或两个以上相互独立控制的区域组成的相互连接的电力系统中，除了控制频率以外，对各个区域的发电量也要进行控制，以便在各个区域之间进行发电计划的调节。对发电量和频率所进行的控制，通常称之为负荷频率控制（LFC）。

电力系统功率和频率控制的方法主要包括：

（1）发电机的调速器控制

（2）额定频率以下卸载

（3）发电自动化控制（AGC）

在大型的电力系统中，发电自动化控制是一种行之有效的功率和频率的控制方式。在相互连接的电力系统中，发电自动化控制的首要目标有两个，一是将频率调节到指定的额定值，二是通过调节所选定的发电机的输出功率，维持各控制区域之间的计划发电量之间的交换。该功能通常就称之为负荷—频率控制。发电自动化控制的第二个目标是在发电机组之间分配所需的发电量的变更，以使运行成本最小化。

无功功率与电压控制

为了能使电力系统可靠且高效运行，应对无功功率和电压实施控制，以达到以下目的。

（1）系统中所有设备的电压极限应在一个所允许的范围之内。供电部门的电力设备和用户的电力设备都设计为在某一固定的额定电压下运行。设备长时间的过电压运行，可能对运行性能产生负面作用，且有可能导致设备的损坏。

（2）最大化地利用输电系统的容量可增加系统的稳定性。

（3）无功功率涌流最小化，以便将设备和输电线路的损耗减小至实际可行之最小值。这就确保了输电系统的有效运行，也就是说确保了有功功率的传输。

将电压维持在必需范围内的难题，又被电力系统中的现实复杂化了，这个现实就是系统为其提供电能负载的数量不计其数，且为其供电的发电机组的数量也是很大的。当负荷变化时，输电系统的无功功率的需求也是变化的。由于无功功率不能长距离传输，因而电压控制的功能，就不得不采用在整个系统中分散布局的某种专用装置来完成。这就与频率控制时的情形截然相反，频率控制是取决于系统总的有功平衡的。恰当地选择和布局控制无功功率和电压的装置，是电力系统工程所面临的主要挑战之一。

电压的控制是由对电能的生产、吸收和系统中各个层面上的无功功率涌流的控制来实现的。发电机组提供了电压控制的基本形式，其自动电压调节器控制着励磁电流，以将发电机的端电压维持在某一预定值。要控制整个系统中的电压，还有一些其他的方法也是经常采用的。运用于电压控制目的的装置可分为以下几类：

（1）无功功率源，例如串联电容器、并联电抗器、同步调相机及静止无功补偿器。

（2）线路电抗补偿器，例如串联电容器。

（3）可调压变压器，例如分接头可调变压器和升压器。

同步调相机和静止无功补偿器提供有源补偿，其吸收或发出的无功功率是自动调节的，以便将其所连接的母线电压维持稳定。

Unit 3.3　Relaying Protection in Power System

3.3.1　Text

Relays or protection systems play a very important role in the sequence of events that lead to power system blackouts. Insecure or failed protection systems, which remain undetected, can make a bad situation worse. It should be clear that every type of relay failure does not fall into this category 【1】. Failures that lead to a misoperation immediately, and can be detected and corrected right away, are not of interest for our present discussion. We are interested in failures that are not activated until some abnormal states of the power system are reached 【2】. These abnormal states are, typically, faults, overloads, reverse power flows, etc. It is known that under-detected (or hidden) failures in protection systems commonly lead to multiple contingencies, which in turn can lead to power system black outs.

The operation of all commonly applied protective relay schemes on HV and EHV systems have been analyzed from the point of view of failure. The application of the fundamental protective principles of zones of protection, selectivity and coordination has been reviewed. Specifically, the use of communication channels for directional comparison and phase comparison relays and transfer trip schemes have been examined from the point of view of possible failures. These failures have been further examined to determine if hidden failures (HF), that is, failures that do not make themselves immediately known, play an important role in extending the disturbance. If a fault occurs in an area that is relatively close to the location of the relay with the HF, its effects may reach the relay. In other words, each hidden failure has a region of vulnerability (RV) associated with it. The RV is an area within the reach or setting of a relay or relay scheme which would cause it to mis-operate if other supervising or control parameters did not prevent its operation. To assign a numerical value to this combination of HFs and RVs and allow the comparison of the various scenarios, a vulnerability index associated with the relays or relay schemes when they are subjected to some contingency was computed. The contingency used to have two lines removed from service, one line due to a fault and the other due to a HF. The study of rare events in a large system presents a formidable amount of calculations with increasing doubt as to the reliability of the overall long-term behavior. To overcome this difficulty, using the technique of "importance sampling" is used to make this problem more manageable. In this technique, the probabilities are altered so that the rare event happens more frequently and the effect of reducing the probabilities can be analyzed. With knowledge of the degree of severity of HF or the probability of the occurrence of mis-operations, the possibility of monitoring the protection and thereby correcting the situation is discussed.

In most cases, the initiating event was some natural event or device failure. The subsequent relay system misoperation then contributed to the disturbance becoming major. In other words, a "hidden defect" in the relay which was not seen prior to the disturbance, caused the relay or a controlling element of the relay to mis-operate or fail to operate because of the stress of the disturbance.

A hidden failure is defined to be a permanent detect that will cause a relay or a relay system to incorrectly and inappropriately remove a circuit element (s) as a direct consequence of another switching event. The defect must be capable of being monitored. A failure that results in an immediate trip without any prior event is not considered a hidden failure. The power system must be planned and operated to withstand the loss of any element.

Hardware failures that result in the relays failing to operate a breaker and trip out a faulted line or device is also not considered a hidden failure since redundant or backup protection or breaker failure systems must normally be provided for such a contingency. A defect or malfunction that occurs at the instant of a fault or switching event, e.g. a "hole" in the blocking signal or an insulation failure caused by a switching surge, was similarly not considered a hidden failure since such a failure is not a permanent condition and cannot be monitored or detected before hand.

The possible failure modes of each relaying scheme leading to vulnerable states of the power system were studied. However, not all of the modes lead to a hidden failure. The hidden failure can be exposed by one or several abnormal power system states. In other words, each hidden failure has a region of vulnerability associated with it. If the abnormal power system state occurs inside the region of vulnerability, the hidden failure will cause the relay to incorrectly remove the circuit element.

There are two types of input signals received by the relaying schemes that have been examined for regions of vulnerability.
　—Distance (impedance or reactance).
　—Current magnitude.

The current magnitude relays operate for the input current magnitude whenever the current exceeds their pick-up setting. The idea of relays with a hidden having a region of vulnerability surrounding themselves may not be directly applicable to these relays. If, however, the input signal is a fault current, we can indirectly apply it. This is, in general, due to the fact that the further the fault is from the relay, the smaller is the fault current. Consequently, we can find the locations beyond which the relay would not operate. Those locations will form a boundary of the RV for current magnitude relays.

In spite of the obvious importance to the industry, there has been little analytic or simulation work in the area of cascading outages of the bulk power system. The reasons are based on the enormous complexity of the problem. The difficulty of simulating rare events coupled with the lack of data on which to base models, have precluded work in the area. The probability of a simple event leading to a widespread outage is greatly reduced by self-checking and monitoring of individual relays. A technique has been established for determining the probability of a cascading outage in terms of the individual probability of a hidden failure in a relay leading to an incorrect trip. The conclusion is not dependent on the exact numbers used for the probability that an exposed line will trip that we treat the individual probabilities parametrically. This technique is referred to as "importance sampling". A significant result of this analysis is that, in a tightly interconnected system, the degree of improvement in system performance is greater than the required degree of improvement in individual relay systems through self-monitoring and control.

If the protection system is essentially digital, it is reasonable to expect different failure modes and different probabilities of hidden failures from the protection system. It is anticipated that digital systems with there inherent ability (via software) to perform self monitoring will allow them to be more reliable than analog systems. Future studies are necessary to verify these beliefs and develop the appropriate self-checking, self-calibrating and control strategies for digital systems. Given the existing analog systems, however, it is possible to imagine intervention by a rational control scheme implemented in a microprocessor-based system to improve performance. The studies of regions of vulnerability and to a lesser extent, the importance sampling work have indicated which relays are the ones with potential hidden failures that can produce the most serious consequences 【3】. Thus, relays with a high-vulnerability index can be identified and their hidden failure modes examined. With the knowledge of which relays are the most vulnerable, counter measures can be taken to reduce or eliminate the likelihood of the hidden failure of key relays.

A computer-based monitoring and control system can be placed in the substation to control the critical relays which have a high-index of vulnerability index, and their vulnerable functions have been identified. The Hidden Failure Monitoring and Control System (HFMCS) has as inputs the same signals which go to the high vulnerability relays. The outputs of HFMCS are combined logically with the outputs of the traditional relays in an appropriate logical connection. For example, consider the possibility that the hidden failure is the loss of directionality in an overcurrent relay【4】. The HFMCS would receive the operating and polarizing signals used by traditional relay. It would perform a directional calculation, and provide an output contact. The output of HFMCS would be connected in series with the directional contact of the traditional relay. Commonly-used analog protective systems have hidden failures associated with the following relay characteristics: fault detectors, directionality, distance measurement, timers, and CT and PT inputs. Using the techniques described above, we can identify those lines and terminals that present the greatest hazard to system reliability. We can therefore add to those lines appropriate digital equipment to provide the necessary supervision.

Relay hidden failures are found to be one of the principal causes of major power system failures. An industry survey was conducted to establish causes of power system blackouts. Research was carried out on how to make their occurrence less likely. The results of the research and conclusions drawn from this study can be summarized as follows:

(1) The analysis of the industry survey and various reports of outages from U.S., Canadian and European relay engineers indicates that hidden relay defects can be a major factor causing otherwise routine relay operations to develop into widespread outages.

(2) A new concept of regions of vulnerability of a protection system has been introduced and quantified. A region of vulnerability is an area within a hidden failure may cause the loss of additional system elements. The technique of finding the region of vulnerability and its use has been demonstrated in the report for a representative power system.

(3) Recognizing that probability is an important element of relay operations, and particularly

of relay failures, the technique of "importance sampling" has been used to calculate the relative probability of having hidden failures in order to take appropriate corrective action. It is clear that adaptive relaying, particularly where digital relays are involved, offers such an opportunity before a major disaster occurs. The ability to self-check and self-monitor, coupled with the adaptive features of changing settings or revising control or trip logic offers a potential solution to mitigating or preventing failures from causing cascading outages【5】.

3.3.2 Specialized English Words

    protection system　　保护系统
    power system black out　　系统断电
    HV and EHV systems　　高压和超高压系统
    directional comparison relay　　方向比较继电器
    transfer trip scheme　　远方跳闸方式
    reach or settings of a relay　　继电器的保护或整定（范围）
    rare event　　小概率事件
    probability　　概率
    redundant　　冗余（设备）
    current magnitude relay　　电流继电器
    pick-up settings　　启动值
    importance sampling　　重要性采样
    control strategies　　控制策略
    power line carrier　　电力线载波
    Hidden Failure Monitoring and Control System（HFMCS）　　隐匿性故障监控系统
    overcurrent relay　　过电流继电器
    directional contact　　方向性吸合
    CT　　电流互感器
    PT　　电压互感器
    self-monitor　　自我监控
    adaptive feature　　自适应特性
    revising control　　修正控制
    reverse power flow　　逆潮流
    protective relay scheme　　继电保护方式
    zones of protection　　保护范围
    phase comparison relay　　相位比较继电器
    hidden failure（HF）　　隐匿性故障
    region of vulnerability（RV）　　易损区域
    vulnerability index　　易损坏指数
    importance sampling　　重要性采样
    immediate trip　　立即跳闸

backup protection　备用保护
breaker failure　断路器故障
cascading outages　串级停电事故
self-checking　自我检测
self-calibrating　自我校验
high-vulnerability index　高易损指数
high vulnerability relays　高易损继电器
overcurrent relay　过电流继电器
output contact　输出吸合（信号）
operating and polarizing signal　运行和极化信号
adaptive relaying　自适应继电保护
changing setting　改变设定值
trip logic　跳闸逻辑

3.3.3　Notes

【1】It should be clear that every type of relay failure does not fall into this category.

翻译该句时应该注意两点：一是 It should be clear that 应按规范译法翻译成"应该清楚"；二是 every type ...does not...为部分否定，因此应译为"并非每一种"，而不能译成全部否定，即"每一种……都不……"。因此该句应译为："应该清楚，并非每一种继电器故障都属此列"。

【2】We are interested in failures that are not activated until some abnormal states of the power system are reached.

翻译该句时应该注意的是 are not ...until...不要译成"不……一直到……"，那样的译文容易引起误读，而应译成"一直到……后才……"。因此该句应译为："我们所感兴趣的，是那些一直等到电力系统达到不正常状态后才出现的故障"。

【3】The studies of regions of vulnerability and to a lesser extent, the importance sampling work have indicated which relays are the ones with potential hidden failures that can produce the most serious consequences.

在翻译此句时应该首先看出，to a lesser extent 在句中是一个插入语，为了不让其搅乱视线，可先将其暂时移去，这样主谓结构便很清楚了，这就是 The studies of regions of vulnerability and the importance sampling work have indicated，其意为"易损坏区域的研究和重要性的采样工作均已表明"。主谓结构的意思弄清楚了后，再将插入语 to a lesser extent 补充回原处，可见它是进一步说明 the importance sampling work，即进一步说明"重要性采样工作"的，其原意为"将范围说得更小一些"。此处转译为"或者再将其范围讲得更具体一些"，并将其表示在括号内，以免破坏主谓结构的完整性。此外还应看出，which relays are the ones...为一宾语从句，其为谓语动词 have indicated 的宾语。只是该宾语从句的引导词 which 是一个疑问代词性质的关系代词，其在从句中的含义是"哪一个"。最后还要看出的是，that can produce the most serious consequences 为一限定性的定语从句，其为 potential hidden failures 的定语。故而整句应译为："易损坏区域的研究和（或者再将其范围讲得更具体一些）重要性采样工作均已查明，哪一个继电器是存在着潜在的隐匿性故障的继电器，而该隐匿性故障是将会产生一系列最为严重后果的"。

【4】For example, consider the possibility that the hidden failure is the loss of directionality in an overcurrent relay.

在翻译该句时，首先应该看出其为一个祈使句，祈使动作为 consider，其原意为"考虑"，此处转译为"下面分析"。分析什么呢？the possibility，"可能性"。什么样的可能性呢？那就是由"that"引导的同位语从句 the hidden failure is the loss of directionality in an overcurrent relay，意为"隐匿性故障就是过电流继电器失去方向性"。因此整句应译为"例如，下面分析隐匿性故障就是过电流继电器失去方向性（这个问题）"。

【5】The ability to self-check and self-monitor, coupled with the adaptive features of changing settings or revising control or trip logic offers a potential solution to mitigating or preventing failures from causing cascading outages.

虽然该句很长，但其实为一简单句。其主句为 The ability...offers...，意为"该能力提供了……"。什么样的能力呢？to self-check and self-monitor，"自我检测和自我监控的"能力。光有这些能力就够了吗？不，还应加上 coupled with the adaptive features，"自适应特性"。什么样的自适应特性呢？of changing settings or revising control or trip logic，"改变设定值、或是修正控制逻辑、或是修正跳闸逻辑的"自适应特性。这种能力提供了什么呢？a potential solution to mitigating or preventing failures from causing cascading outages，"为缓解和防止故障而引起的灾难性停电提供了潜在的解决办法"。因此本句应译为："自我检测和自我监控的能力，加上具有改变设定值、或是修正控制逻辑、或是修正跳闸逻辑的自适应特性，可为缓解和防止故障而引起的灾难性停电提供潜在的解决办法"。

3.3.4　Translation

电力系统的继电保护

继电器或保护系统在预防引起电力系统停电的一系列故障中起着非常重要的作用。尚未发现的不可靠或故障的保护系统会使情况更加恶化。应该清楚，并非每一种继电器故障都属此列。立即引起运行不正常的故障，以及可及时检测到并加以纠正的故障，不是此处讨论的兴趣所在。我们所感兴趣的，是那些一直等到电力系统达到不正常状态后才出现的故障。这些不正常状态通常是指故障、过负荷和逆涌流等。众所周知，在保护系统中，那些未检测到（即隐匿）的故障常常引起多种意外事故，进而导致电力系统断电。

从故障的观点出发，对高压和超高压系统中通常运用的继电保护方式的运行，已经进行过分析。对保护范围、保护的选择性及协调性的基本原理的应用进行了回顾。尤其是从可能出现故障的观点出发，审视了方向比较继电器和相位比较继电器的通信通道的运用，也审视了远方跳闸这种操作方式。此外，还进一步排查这些故障，以确定那些隐匿性的故障（即其自身不能立即发现的故障）是否对故障的扩大起了重要的作用。如果故障发生在距安装有隐匿性故障的继电器很近的区域内，该隐匿性故障的影响将波及继电器。换言之，每一种隐匿性故障都有一个相应的易损区域。在继电器或继电器组合所保护的或整定的某区域内，如果其他监测或控制设备未能采取预防措施，该继电器或继电器组合就会产生误操作，该区域就是所谓的易损区域。为了给这一组隐匿故障和易损区域的组合编上号，以便对不同的方案进行比较，就要计算当继电器或继电器组合遭受意外故障时的相关易损指数。通常意外故障有两条线路逃离装置，一条是由于某一种（其他性质的）故障，另一种就是由于隐匿性故障。由于对整个长期运行可靠性的日益担忧，便对大型电力系统中的小概率事件进行了一项研究，结果表明其计算量是相当大的。为了克服这一困难，采用"重要性采样"技术，使得该问题易于解决。在这一技术中，对概率进行了变换，这样小概率事件的发生便变得频繁，因而便可对概率减少的影响进行分析。在已知隐匿性故障的严重程度或不正常运行概率的情况下，讨论对保护进行监控并加以纠正的可能性。

在大多数情况下，初起的一些事故都是一些自然性质的事故或器件故障。而后，由继电器系统的误操作所引起的事故便打乱了这种局面，这种误操作便成了主要因素。换言之，干扰前未曾发现的继电器中的"隐患"，在出现一连串的干扰后，引起了继电器或其中的控制元件的误操作或拒绝动作。

隐匿故障被看成为一种永久性的缺陷，它会使继电器及其系统不正确地或不恰当地切断某一电路元件，从而导致另一开关的误动作。这种缺陷必须能够被置于监控之下。而导致立即跳闸但事先没有故障发生的情况不属于隐匿故障。在制订电力系统的运行计划时，就必须考虑到可能有某些元件会出现故障这一因素。

由于在继电器无法操作断路器，或者无法跳开故障线路及装置这样的紧急事故中，通常冗余设备、备用保护或断路器故障系统都是必须提供的，因此导致的这种硬件故障不属于隐匿性故障。在故障或开关动作瞬间出现的缺陷或功能失灵，即开关冲击引起的闭锁信号中的"空洞"或绝缘故障，同样也不属于隐匿故障，因为这种故障状态不是永久性的，并且预先不能监测或检测得到。

对产生电力系统易损坏状态的每一种继电保护方案中可能出现的故障的形式进行了研究。然而，并不是每一种情况都产生隐匿性故障。隐匿性故障可以由一个或几个不正常的电力系统状态而暴露出来。换言之，每一种隐匿性故障有其相应的易损坏区域。若电力系统在此区域内出现不正常状态，隐匿性故障将引起继电器不正确动作而误切断线路中的元件。

输送给已检测过易损坏区的继电保护方案的输入信号有两类：（1）距离（阻抗或电抗）；（2）电流幅值。只要输入的电流值大于设定的启动值，电流继电器就会动作。有隐匿性故障的继电器附近，存在易损坏区域的概念不直接适用于这些继电器。然而，若输入信号是故障电流，则可以间接适用。也就是说，一般而言，这是基于这样一个事实，即故障离继电器越远，则故障电流就越小。因此，便可找到继电器不动作以外的位置。这些位置就形成电流继电器的易损坏区域的边界。

尽管对电力工业有着明显的重要性，但对于整个电力系统的串级停电事故很少进行分析和模拟的工作。原因是问题的异常复杂性、模拟小概率事件的困难与缺乏形成模拟的数据交织在一起，阻碍了该领域的工作的进展。由于实行了自检和对各个继电器分别进行了监控，因而某一简单事件造成大范围停运的可能性大为减少。已经出现了一种技术，可以根据误动作继电器的隐匿性故障的各概率，确定串级停电事故的概率。该结论与用于计算暴露隐匿性故障线路断开的概率时的确切数目无关，计算时我们将各个概率仅作为参数。这项技术称为"重要性采样"。这种分析的显著成果是得出了这样的结论，即在一个联系密切的系统中，通过采用自监测和自动控制，系统性能的提高程度要大于各继电保护系统的改进程度。

如果保护系统基本是数字式的，自然便可预料其故障状态和隐匿性故障概率是不同的。可以预期，数字式系统因其自身通过软件，具备了固有的自我监测的能力，故其可比模拟式系统更加可靠。为了验证这些看法，以及开发适用的数字系统自我检测、自我校验和控制策略，有必要进行进一步的研究。然而，对于某一个给定的现有模拟式系统来说，可以设想其可用一种，编制在以微处理器为基础的控制系统中的合理控制方案进行干预，以改进其性能。易损坏区域的研究和（或者再将其范围讲得更具体一些）重要性采样工作均已查明，哪一个继电器是存在着潜在的隐匿性故障的继电器，而该隐匿性故障是将会产生一系列最为严重后果的。由此便可确定具有高易损指数的继电器，并且检查其隐匿故障状态。了解了哪些继电器是最易损的，便可采取相应的对策以减少或消除关键的继电器可能存在的隐匿性故障。

一种计算机监控系统可以用于装置在变电所内，以控制那些易损指数高的高危继电器，这种监控系统的价值已经得到了肯定。隐匿性故障监控系统（HFMCS）给易损继电器输入一个与其输入相同的信号。隐匿性故障监控系统的逻辑输出信号，以某种适当的逻辑连接方式，与传统的继电器的逻辑输出信号相叠加。例如，下面分析的隐匿性故障就是过电流继电器失去方向性（这个问题）。隐匿性故障监控系统会接收到传统的继电器所采用的运行和极化信号，其会进行方向性的计算，并输出一个吸合（信号）。隐匿性故障监控系统的输出信号会与传统的继电器的方向性吸合信号相串联。通常所采用的模拟式保护系统，均具有与以下继电器性能相关的隐匿性故障：故障检测、方向判断、距离测量、计时功能，以及电流互感器的输入和电压互感器的输入。采用以上所述的技术，便可确定那些给系统可靠性带来最大危害的线路和终端。这样也就可以在这些线上增加合适的数字式设备，以提供必需的监测。

现已发现，大部分电力系统故障停运的主要原因之一就是继电器的隐匿性故障。已开展一项工业性的研究，以便弄清电力系统停运的原因，以及如何尽可能地减少停运故障。现将研究结果及其从中得出的结论归纳如下：

（1）该份工业调查的分析，以及来自美国、加拿大和欧洲继电保护工程师们的停电报告都得出同样的结论：继电器的隐匿性故障是引起其他的常规继电器动作进而发展为大范围停电的主要因素。

（2）电力系统出现了一个新概念，即保护系统的易损区域，并对其进行了量化。易损区域是一个区域，在该区域内隐匿性故障可能引起加入系统元件的丢失。确定这一区域的技术及其用途，已经在一个典型的电力系统的报告中展示。

（3）认识到这种概率是继电器运行的一个重要组成部分，尤其是对于继电器的故障来说，"重要性采样"技术已经用于隐匿性故障的相对概率的计算之中，以便采取相应的纠正措施。显然，自适应继电保护尤其是采用了数字式继电保护后，可以在产生严重危害前提供一个机会。自我检测和自我监控的能力，加上具有改变设定值、或是修正控制逻辑、或是修正跳闸逻辑的自适应特性，可为缓解和防止故障而引起的灾难性停电提供了潜在的解决办法。

Unit 3.4　SCADA & EMS in Power System

3.4.1　Text

At the heart of most computer programs lies a model of the problem domain. This model defines the entities which the program is designed to handle and describes their characteristics and state. For non-trivial software systems it is considered for good practice to separate functionally the model from the applications which analyze or modify this model. Such explicit models can be easily extracted and stored in databases.

A SCADA system requires model of the equipments used to monitor and control the power system. The SCADA model is suitable for the data acquisition and supervisory tasks. When security and scheduling applications were added to SCADA systems to created Energy Management System (EMS), a model of the power devices was added to the SCADA model. While these two models describe different aspects of the same entity (the power system) they are often kept separate for practical and historical reasons.

AI(Artificial Intelligence)applications, like the conventional advanced applications(dispatcher power flow, security analysis, optimum power flow), need a model of the power system. Most of the information needed by the AI applications can usually be extracted from the model used by the advanced applications. However, this model is usually organized in indexed tables, a type of structure which facilities the development of high-performance algorithmic programs but which is not practical for AI applications because they are usually built using specialized tools or shells. The power, flexibility and performance of these shells can only be fully achieved if the data is organized in specialized data structures within the shell. AI applications therefore require their own model of the power system.

Adding another power system model to the EMS involves the implementation of mechanisms to populate the model from a database, to initialize the dynamic attributes of this model from the real-time data, and for updating these attributes when a change takes place in the power system. It could be argued that the need for a separate model is caused by the AI shell and that creating a specialized AI shell working directly from the conventional data structures would solve this and other problems 【1】. While theoretically possible, such a tool would not have the flexibility and power of general purpose tools and would seriously hamper the development of flexible AI applications. Since no such tool is commercially available, it would have to be custom made and proprietary. Developing such a tool investment and its adoption would go against the current philosophy of EMS design practices.

As their name suggests, shells tend to isolate the AI applications from the rest of the EMS. While the shells currently available on the market are considerably more open than tools which were available a few years ago, a special interface must still be built to integrate the AI applications with the other EMS subsystems. In particular, this interface should support the following interactions:

Real-time Data: AI applications must base their reasoning on the actual state of the power system. It is thus essential that they be informed of the significant event taking place in the power system and that they have the ability to obtain the value of specific telemetered data.

Relational Database: Most modern energy management systems include a relational database which is used to store the primitive data description of the system as well as historical data. AI applications need to have access to this database so they can build their model and, if necessary, store some of their results.

User Interface: The conclusions and other results produced by the AI application should be displayed using the standard user interface of the EMS. This requirement is particularly important if one-line diagrams and other graphical displays are used to present these results.

Numerical Applications: The conventional numerical applications of an EMS (power flow, security analysis, etc.) can provide information which can greatly enhance the power of AI applications. AI applications should therefore have the ability to setup and run numerical applications and analyze their results.

Supervisory Control: In some situations it may be desirable to allow an AI application to issue commands through the supervisory control subsystem.

Configuration Control: It should be possible to startup, shutdown and check the status of the AI applications through the standard configuration control mechanisms.

Creating a power system model and a set of interfaces for each AI application increases considerably the development and integration costs of these applications. In order to reduce the costs, the idea arose to create an environment into which various AI applications could be plugged easily. This environment would provide all the interfaces with the EMS required by the applications. It would also support a model of power system shared by all the AI applications. Since the various AI applications would reside in the same environment, inter-application communication mechanisms could easily be setup. This environment would therefore facilitate the development of AI applications capable of collaborating on the solution of related problems. An environment which supports the modeling and interfacing requirements outlined in the previous section has been designed and built. This environment is called ODE (Operate Decision Environment).

The architecture of ODE is based on a layered concept similar to the one used for the specification of computer systems. Four layers may be identified as follows:

(1) The Data Layer: containing all the static data, knowledge bases, and information needed to support the reasoning and decision making activities at the higher layers.

(2) The Data Access Layer: provides the needed mechanisms to access information from the Data Layer. In addition, it provides a representation media for data and knowledge used by the various AI applications residing in the upper layers. The data access layer also includes interfaces to the EMS for exchanging of data with the various software components comparing the traditional side of the EMS (in particular the SCADA system).

(3) The Application /Tool Layer: contains all the AI and ES applications, as well as, the

various AI and ES tools used with these applications. The tools considered here are tools that support reasoning and problem solving in ES and, more generally, in AI problem solving. These tools may be used in applications ranging from Diagnosis, to planning, to design.

(4) The Decision Layer: consists of a knowledge-based component that oversees the orchestrated execution of all the tools and applications in the Application/Tool Layer, and their relationship to the numerically intensive applications. The Decision Layer, just as the Data Access Layer, is connected to the EMS to access and provide the required decision and control information for reasoning tasks.

In the early stages of its design, it was decided that ODE should facilitate the integration of the applications but should not put too many constraints on their development. In other words, each application should be able to take advantages of the services provided by ODE without having to be completely folded into ODE. This decision to preserve the independence and the modularity of the applications led quite naturally to the adoption of a client/server implementation architecture that contains elements of all four of the architectural layers.

ODE consists of independent processes which can be divided in three types:

(1) Data sever: the OdeRT process is the heart of the system. It supports the model of the power system and acts as a focal point for communications among the applications and with the rest of the EMS.

(2) Interfaces: OdePOP is responsible for the model of the power system while OdeCOM supports all communications between ODE and the rest of the EMS.

(3) Applications: Processes responsible for intelligent alarm processing (IAP), fault diagnosis (DFS) and restoration assistance have been integrated within ODE. Other applications (such as a switching assistant) will be integrated into ODE in near future.

The three applications which have been integrated within ODE share a need for topological information:

(1) IAP analyzes changes in network topology to determine whether they correspond to an abnormal situation (such as a bus spilt or the creation of an island) which must be brought to the attention of the operator.

(2) DFS starts its in diagnosis of a fault by making a list of all the components included in the blacked-out area.

(3) RA needs to know which parts of the system are energized and which ones are de-energized so it can determine what restoration actions can be taken and how they can be carried out safely.

All this information is obtained from the common model of the power system maintained by OdeRT. This form of pooling is not limited to topology: if two or more applications have similar requirements, the necessary functionality can be implemented in OdeRT instead of being duplicated in the applications. Maintaining the topological model of the power system is one of the tasks in OdeRT.

Since AI applications which have been integrated into ODE are designed to help the operator make decisions in real time, the model should track the evaluation of the system with a minimum of delay. In order to meet this performance requirement, updating the model must be performed incrementally after each significant event.

In addition to supporting the initialization and the updating of the power system model, the process OdeCOM is designed to provide a full, tow-way communication channel between ODE and the other subsystems of the EMS. In particular, OdeCOM provides access to the user interface and supervisory control subsystems of the EMS as well as to the conventional numerical applications. To accomplish this, OdeCOM implements an interface between OdeRT and the message bus of the EMS.

When an ODE application needs to request a service from another EMS subsystem or needs to send information to another EMS subsystem, it creates a member of a class operation. This object is passed to OdeCOM which constructs a message based on the data encapsulated in the object and dispatches it to the appropriate address. Conversely, when OdeCOM receives a message from another EMS subsystem, it creates a member of the class event, sets the attributes of this new object based on the information contained in the message and passes it to OdeRT for actual processing.

It should be noted that OdeCOM translates the names which are used by ODE to refer to components of the power system into the address used by the other EMS subsystem and vice versa.

Integrating expert systems and AI applications in an EMS environment in a cost-effective manner is not an easy task. This paper has discussed the issues which must be addressed when designing this integration. It has been argued that these issues should be resolved by creating an environment which supports all the interfaces with the EMS which the AI applications will need. This environment should also be responsible for maintaining a model of the power system common to all the AI applications. Once this environment has been created, AI applications can be easily "plugged" into the EMS.

This streamlined integration process not only lowers the cost of installing and maintaining an existing application in an EMS but also considerably reduces the cost of developing new applications.

The common power system model is a very powerful medium for interactions among applications. While little work has yet been done in this area, it appears that there is a considerable potential for the development of collaborative.

3.4.2 Specialized English Words

non-trivial software system 重要的软件系统
Energy Management System（EMS） 能量管理系统
dispatcher power flow 功率涌流调度
optimum power flow 功率涌流优化
telemetered data 遥测数据
one-line diagrams 单线图
fault diagnosis（DFS） 故障诊断
restoration assistance（RA） 辅助修复

minimum of delay　最小时延
data file　数据串，数据单
cost-effective manner　低成本高收益方式
incremental topology processor　增量拓扑信息处理机
standard configuration control mechanism　标准的框架控制机构
layered concept　层化概念
data layer　数据层
data access layer　数据通道层
fault diagnosis（DFS）　故障诊断
switching assistant　开关协助
expert system　专家系统
SCADA（Supervisory Control And Data Acquisition）　监控与数据采集系统
AI（Artificial Intelligence）　人工智能
security analysis　安全分析
high-performance algorithmic program　高性能的计算程序
intelligent alarm processing（IAP）　智能报警系统
knowledge-based switching advisor　以知识为基础的开关顾问
two-way communication channel　双向通信通道
streamlined integration process　流线型的集成过程
object-oriented　面向对象的
supervisory control subsystem　监控分系统
Operate Decision Environment（ODE）　运行决策环境
reasoning and decision　论证与决策
intelligent alarm processing（IAP）　智能警报处理
restoration assistance　复归协助
message bus　信息母线

3.4.3　Notes

【1】It could be argued that the need for a separate model is caused by the AI shell and that creating a specialized AI shell working directly from the conventional data structures would solve this and other problems.

该句的句干很简单，It could be argued，意为"这一点也许（译者注：此处用的是虚拟语气，也就是说其实并不见得会起争论）会引起争论"，哪一点呢？就是由 that 引导的主语从句 the need for a separate model is caused by the AI shell and that creating a specialized AI shell working directly from the conventional data structures would solve this and other problems，意为"（既然）另外还需要一个不同的模型是由人工智能外壳引起的，那就直接生成一个不同于以往的数据结构人工智能外壳，不就解决这个问题吗？其他问题也应该不在话下呀。"但若将 be argued 直译成"引起争论"，会引起歧义，产生误解。其实此处 be argued 并不是真正的"争论或争吵"，而是"提出疑问"。因此该句应译为："也许有人会问，既然另外还需要一个不同的模型是由人工智能外壳引起的,那就直接生成一个不同于以往的数据结构人工智能外壳,不就解决这个问题了吗?其他问题也应该不在话下"。

3.4.4 Translation

电力系统中的监控和数据采集系统与能量管理系统

  大部分计算机程序的核心问题就在于研究问题的模型。而这一模型也就决定了,那个设计用来处理和描述这些问题的特征和状态的程序的所有一切内容。对于重要的软件系统,为了便于实用,应该考虑将理论上的模型与其实际应用区别开来,实际的应用又是可以反过来分析和修改模型的。这样一种表达明晰的模型可以很容易在数据库中提取和存储。

  在运用监控与数据采集系统(SCADA)时,需要用于监测和控制电力系统的设备模型。这种 SCADA 模型只适应于数据采集和监管的场合,而为了生成能量管理系统(EMS),在 SCADA 系统中增加了安全性和计划表应用时,就要将电力装置的模型加入 SCADA 中。尽管这两种模型都是描述电力系统这同一个对象的(只是不同侧面而已),但由于实际的和历史的原因,这两者又常常是各自独立的。

  人工智能的应用,像通常其他以往的先进技术(如功率涌流调度、安全性分析、功率涌流优化等)的应用一样,是需要电力系统模型的。AI 应用时所需的大部分信息通常都可以从以往的先进技术应用时的模型中提取。然而,这种模型常常是编制在索引表中的,采用这种形式的数据结构,为编制高性能的计算程序提供了便利,但由于其通常是采用一种专门的工具(或称为外壳)建立起来的,因而不适合于 AI 的应用。这些外壳的功能、柔性和特性只有在将数据按照专门的数据结构编排在外壳内时才能得到体现,因此 AI 的应用需要电力系统本身的模型。

  要将另外一个电力系统模型加入能量管理系统,就会涉及以下几个过程,即将模型从数据库中迁入,根据实时数据将该模型的动态特征初始化,当电力系统发生变化时适时地更新这些特征。也许有人会问,既然另外还需要一个不同的模型是由人工智能外壳引起的,那就直接生成一个不同于以往的数据结构人工智能外壳,不就解决这个问题了吗?其他问题也应该不在话下。尽管这在理论上是可行的,但这样的一种工具不仅不具备通用工具所具备的柔性和功能,还会严重影响人工智能应用的柔性。由于这样的工具在市场上还买不到,因此只能量身订制和专卖。研制这样的工具需要大量的投资,推广它的应用也是与能量管理系统当前设计的实际情形背道而驰的。

  顾名思义,外壳就是有一种将人工智能的应用与能量管理系统的其他部分相隔离的作用。尽管目前市场上可买到的外壳比几年前所受的限制要少些,但若要将人工智能应用与能量管理系统的其他分系统结合起来,仍然必须设置一种特殊的界面。具体说来,这种界面应能支持以下几方面的功能。

  实时数据:人工智能应用的依据必须以电力系统的实际状态为基础,因此了解电力系统中发生的重大事件,以及具备获得明确的遥测数据的能力,对人工智能的应用来说是最为基本的。

  相关的数据库:大多数能量管理系统都有相关的数据库,其用于存储描述系统的原始数据和历史数据。人工智能的应用应能设置进入数据库的通道,以便建立其系统的模型,必要时还可存储一些运行结果。

用户界面：人工智能的应用所产生的结论和其他一些运行结果，应该运用能量管理系统的标准用户界面来显示。特别是在运用单线图和其他图形显示来表示这些结果时，这一需要尤为重要。

数字应用：能量管理系统以往的数字应用（如功率涌流、安全性分析等）都能提供大大增强人工智能的应用功能的信息。因而，人工智能的应用应该有能力建立和运行数字式应用，并分析其结果。

监控：在某些情形下，可能希望人工智能的应用可通过监控分系统向下一级系统发布命令。

框架控制：应该可通过标准的框架控制机构来启动、关闭以及检查人工智能应用的状况。

对每一个人工智能的应用，都要建立电力系统的模型和一系列与之相关的界面，这将大大地增加其开发和实施的费用。为了减少这些费用，便出现了一个设想，即建立某种可将各种不同的人工智能应用，都能很容易地囊括其中的环境。运用这种环境，便可根据能量管理系统的需要，为其提供所有的界面。这样，其对电力系统模型与所有的其他人工智能应用共享也是支持的。由于不同的人工智能应用处在同一环境中，因此相互应用的通信机制便可很容易地建立起来。因而，这一环境提高了人工智能应用在解决相关问题时的协同能力。一种能够支持如前一节所述的建模和设置界面需求的那种环境已经设计和建成了，其称之为运行决策环境（ODE）。

运行决策环境的结构建立在层化概念的基础上，该概念类似于在计算机系统中的运用，其按定义可分为如下4层。

（1）数据层：包括所有的静态数据、知识库，以及支持更高层进行论证与决策活动所需的信息。

（2）数据访问层：提供通向数据层获取所需信息的途径。此外，其还为驻留在其他几层的各种在人工智能应用中所采用的信息和知识，提供展示的媒介。数据通道层还含有与能量管理系统提供界面，以便于其与能量管理系统的传统侧的各种不同的软件间进行数据交换（尤其是监控与数据采集系统，更是如此）。

（3）应用/工具层：含有人工智能和能量管理系统的所有应用，及其在这些应用中所使用的工具。此处所说的工具，是指支持在能量管理系统的应用中探究和解决问题的工具，或者更一般地说，是在人工智能的应用中解决问题的工具。这些工具的运用范围可从诊断到规划，又从规划直至设计。

（4）决策层：含有一个以知识为基础的元件，该元件监督在应用/工具层中所有工具的编制和应用的情况，以及与它们的进一步运用之间的数量关系。决策层与数据层一样与能量管理系统相连，以提取和提供探究任务时所需的决策和控制信息。

在设计的前期阶段，必须明确运行决策环境应为各种应用的集成提供便利，但不应过多地限制它们的生成。换言之，每种应用均可享用运行决策环境所提供的各种服务，而又不必完全束缚于其中。这种既可以保持独立性，又可以在应用中进行修改的决策方案，自然而然地成了客户和供应商都能接受的组成结构，该结构包括了所有的4层结构。

运行决策环境含有几个独立的处理功能，其可分为以下三种。

（1）数据服务：OdeRT 处理功能是系统的核心部分，其提供电力系统的模型，并在各种应用与能量管理系统的其他部分进行通信时起到了焦点作用。

（2）接口：OdePOP 对电力系统的模型作出响应，而 OdeCOM 则为运行决策环境与能量管理系统的其他部分提供通信联系。

（3）应用：负责智能警报处理、故障诊断和复归协助的处理功能，均已集成在运行决策环境之中了。其他的应用（如开关协助）在不久的将来也会集成到运行决策环境之中。

以上在运行决策环境中的三种应用均需共享以下拓扑信息。

（1）智能警报处理：其分析网络拓扑的变化，以确定是否发生了不正常的情况（例如，母线断裂或产生了孤立的地区），这些都是必须引起运行人员注意的事件。

（2）故障诊断：其通过制作出停电区域内所有元件的清单，着手开始其对故障的诊断。

（3）复归协助：其需要了解系统的哪个部分处在供电状态，哪个部分处在非供电状态，以便能确定采取什么样的恢复措施，并如何安全地得以实施。

所有这些信息，都是从保存在 OdeRT 中的电力系统公共模型中得到的。这种资源共享的形式，不仅仅限于网络的拓扑结构。如果两个或更多的应用的要求是相同的，则所需的功能可以设置到 OdeRT 中，而不是复制到其应用中。保存电力系统的拓扑模型是 OdeRT 的任务之一。

由于结合在运行决策环境的人工智能应用，是用来帮助操作人员作出实时决定的，因此系统的模型必须紧跟电力系统的变化（只允许有很小的一点延时）。为了满足这种性能要求，就必须在每一重大事件后不间断地更新系统模型。

除了提供电力系统的原始模型和更新后的模型，OdeCOM 处理功能是设计用来为运行决策环境与能量管理系统的其他分系统，提供全面的双向通信通道的。具体地说，OdeCOM 提供进入用户界面和能量管理系统的监控分系统的通道，同时也为常规的数字式应用提供通道。为了实现这一功能，OdeCOM 在 OdeRT 与能量管理系统的信息母线之间设置了界面。

当运行决策环境需要从能量管理系统的另一个分系统得到服务，或是需要给能量管理系统的另一个分系统发送信息时，其便进行若干某一级别的运算。该运算结果传送到 OdeCOM，在此以封存在其内的数据为基础形成一组信息，并将其发送至适当的地址处。反过来也是类似的，当 OdeCOM 接收到从能量管理系统的另一个分系统来的信息时，其也进行若干某一级别的运算，并将基于这些信息的最新运算结果的内容设置成一组信息，从而传送到 OdeRT，以便进行实际的处理。

应该注意的是，OdeCOM 对照着电力系统的各个元件，将在决策环境中运用的名称转换为能量管理系统的其他分系统所采用的地址，反之亦然。

将专家系统与在能量管理系统环境下的其他人工智能的应用，集成为一个低成本高收益的集成系统并非易事。本文讨论了设计这一集成系统时必须要涉及的问题。有人曾对这些问题提出过质疑，即解决的方法应该是通过生成一种环境来加以解决，这种环境应能提供在人工智能的应用中所需的与能量管理系统相关的所有界面。该环境也应能负责维护，即所有通用于人工智能应用中的电力系统模型。一旦产生出这样的环境，人工智能的应用便可以很容易地嵌入能量管理系统之中。

该流程性的集成过程，不仅可以降低现有的能量管理系统应用的安装和维护费用，而且还可大大地削减开发新应用的成本。

通用的电力系统模型，在各种应用的相互作用中是功能很强的媒介。由于这一领域内的工作做得甚少，因此看起来在其协调发展方面，似乎还有可观的研究潜力。

Unit 3.5   Electric Power Flow Calculation in Power System

3.5.1   Text

Distribution Power Flow Calculation Method Outline

At present, the traditional power flow calculation methods, such as Newton - Raphson method, PQ decomposition, are targeted at high-voltage power grids; and in the distribution network of lower voltage, the circuit characteristics and load characteristics are related to high-voltage power grids have very different, making it difficult to direct application of the traditional power flow calculation 【1】. Due to the lack of effective, for a long time the electricity sector calculation in the distribution of power flow was with the majority of hand-counting method. Since the early 80's, the domestic and foreign experts and scholars have made a development of a variety of distribution power flow computer algorithms based on the hand count method. At present, the trend of the calculation in the radiation-type distribution network is mainly in the following two categories:

(1) Direct application of Kirchhoff Voltage and Current Law

First of all, compute the current into nodes, then solve the branch current, finally solve the node voltage, and take the power error at network nodes as a convergence criterion, with such as the slip algorithm, the voltage/current iteration less mesh distribution power flow algorithm and the direct method, such as circuit analysis 【2】.

(2) With Newton - Raphson method for solving the equation of state

Take the active power P, reactive power Q and the square of the node voltage as the state variables in the system, write out the state system equation and use Newton - Raphson method to solving the equation, which can directly get the system power flow solution.

Distribution Network Power Flow Calculation Difficulties

Data collection

In the distribution network flow calculation, the network data and run data integrity and accuracy are the major impact factor in calculation accuracy. For the actual operation sector, to provide the complete and accurate network data and operational data of the distribution network is difficult, which are mainly the following reasons:

(1) Due to the complexity of the network structure of the distribution network, especially in the voltage 10kV and below distribution network, and multi distributed users, it is not possible to install measuring meters in every distribution feeder and branch, which makes it difficult for the run departments to provide complete and accurate operating data.

(2) In the actual distribution system, some main lines have the automatic measuring table

installation, but most of the distribution network can only be with the manual data collection, so it is difficult to ensure the accuracy of the run data, which limits the power flow calculation accuracy, making the majority of results can only serve as reference materials, and can not be used for actual decision-making 【3】.

Load Redistribution

Due to the complexity of the network structure, the user equipments in a wide range types and different from each other, and less measurement in the distribution network, the department can not get the statistics of each distribution transformer load curve and just provide the quite accurate total load curve at the distribution network root node (that is, the low-voltage side exit in a step-down transformer). Thus in the calculation of the trend of distribution network in, it is the key issue of power flow calculation, to take what load distribution method to distribute, as really as possible, the total load at the root to all load nodes.

Practical Flow Calculation Program

In the distribution power flow calculation, when the three-phase distribution system load is balanced, the model of this system may be replaced by the one of the equivalent single-phase system, in which the power distribution line is replaced by the resistance and reactance in each unit length and the capacitance of line parallel capacitor sets (taken as a parallel loads) can be temporarily ignored. For the radiated longer lines, the injected line capacitance volume can be taken as a load capacitance. The load-voltage characteristics of all loads are ignored, and that is to say the load power does not change with the voltage amplitude.

In doing the actual power flow calculation in the distribution network, based on the actual characteristics of distribution networks and the characteristics of the data provided by the electricity section, the following assumptions are made:

(1) It is assumed that the three-phase distribution network is a equilibrium one, which can be calculated as an equivalent single-phase network.

(2) Assume each distribution transformer at load at the same time has the same load factor.

(3) It is assumed that the distances between any two main masts and any two branch masts is fixed, with which the total length of each line could be calculated with the distances between any two masts and the number of total masts.

(4) Assume all load power factors are same.

(5) It is assumed that all distribution transformers are distributed within the total load on the root node, according to its rated capacity, under the same load factor.

To do the power flow calculation based on the above assumptions, there are two practical problems need to solve: how to conduct effectively, accurately and fast the track branch condition and how to distribute the load. As the distribution network is a form of radiation, and the distribution feeders are more complex in structure and the number of branch lines on them are much many, so when doing the calculation, each main distribution feeders and branch feeder lines are regarded as a separate loop and then calculate their the trend value respectively. In calculation, it is assumed that the voltage value at the main feeder root node (step-down transformer bus exit) is constant and the branch root node voltage (ie the voltage at the node on main feeder line) is known and constant in the calculation. In each calculation, the trend of the main feeder lines is first calculated, then the node voltage values are updated (except the one outside the main root feeder), and then the trend of all branches are calculated.

As the power sections can only provide the load curve of distribution feeder root node and can not provide the load curves of each distribution transformer, based on the actual characteristics of distribution network, this paper distributes the total load at the root node according to a certain load distribution coefficient and the rated capacities of all transformers, and the obtained result has a smaller error compared with the actual situation 【4】. But this way has a shortcoming that it does not take full advantage of a small number known load values in at some nodes and if with the state estimation method to estimate the node loads, the advantage of the known node load can be full taken and the calculation accuracy can be improved 【5】.

In the reverse calculation, if there are a number of branches at the root node, the root node voltage value is taken as the average voltage value of all branches connected with it, and in the prior calculation, the total power load at root node is assigned to each branch.

In accordance with the ideal above, the programme is made in this paper and the calculation results meets with the actual, which demonstrates the feasibility of the program.

Digital Simulation

Based on the analysis above, this paper uses DistFlow law, by-slip algorithm and Newton-Raphson method, respectively, to calculate an example of the IEEE standard 69-node and 33 node network power flow calculation. Table 1 (omitted, the following is same) is the comparison of the calculated results in 69-node network with the three methodes, Table 2 is the one in 33-node network results and Table 3, 4, 5 are the analysis results of two lines under the jurisdiction of the Guiyang City Power Supply Bureau, with the by-slip algorithm and DistFlow algorithm.

From the above table we can be see in accordance with the method in this paper, in the case that the branch parameter r/x ratio is larger, there often will be divergent or oscillation phenomenon in the calculation with the standard Newton - Raphson method, because of the own distribution network showing a pathological network characteristics. And if with the by-slip algorithm and DistFlow algorithm, the convergence is better and both the calculation speed and the convergence accuracy are high. However, the slip algorithm is more sensitive to some networks (such as 69-node network) thus in the case of higher convergence accuracy, the maximum error remains a constant value, making the calculation does not converge.And if with DistFlow law, there will not be this phenomenon. But in general, in the same accuracy convergence, the by-slip algorithm converges faster than DistFlow one.

3.5.2 Specialized English Words

computer algorithm　计算机算法
distribution power flow calculation method　配电网潮流计算方法

distribution of power flow 潮流分布
power error 功率误差
current into node 注入节点的电流
node voltage 节点电压
voltage / current iteration 电压／电流迭代法
circuit analysis 回路分析法
distribution network 配电网络
load redistribution 负荷再分配
main mast and branch mast 主线杆和分支线杆
step-down transformer bus exit 降压变压器母线出口
state estimation method 状态估测法
divergent phenomenon 发散现象
electric power flow calculation 电力系统潮流计算
Newton - Raphson method 牛顿-拉夫逊法
PQ decomposition PQ 分解法
radiation-type distribution network 辐射式配电网
convergence criterion 收敛判据
branch current 支路电流
slip algorithm 逐一支路算法
less mesh distribution power flow algorithm 少网孔配电网的潮流算法
data collection 数据采集
distribution feeder and branch 配电馈线和分支线
distribution network root node 配电网络根节点
trend value 潮流值
load distribution coefficient 负荷分配系数
digital simulation 数字仿真

3.5.3 Notes

【1】At present, the traditional power flow calculation methods, such as Newton - Raphson method, PQ decomposition, are targeted at high-voltage power grids; and in the distribution network of lower voltage, the circuit characteristics and load characteristics are related to high-voltage power grids have very different, making it difficult to direct application of the traditional power flow calculation.

虽然该句较长，但并不复杂，还是比较好翻译的，只要看清楚其为两个并列子句组成，因此一是要注意汉译时要译成两句（这样句子较短，符合汉语的习惯）。二是要注意不要将并列连词 and 译成"和"，而应将其转折口气译出来，即译成"而"。因此全句应译为："目前，传统的电力系统潮流计算方法，如牛顿-拉夫逊法、PQ 分解法等，均以高压电网为其运用对象。而配电网的电压等级较低，其线路特性和负荷特性都与高压电网有很大区别，因此很难直接应用传统的电力系统潮流计算方法"。

【2】First of all, compute the current into nodes, then solve the branch current, finally solve the node voltage, and take the power error at network nodes as a convergence criterion, with such as the slip algorithm, the voltage / current iteration less mesh distribution power flow algorithm and the direct method, such as circuit analysis.

该句为一祈使句，祈使的动作有四个 compute、solve（两个）、take。句后的 with such as the slip algorithm, the voltage / current iteration less mesh distribution power flow algorithm and the direct method, such as circuit analysis 为一介词短语，做该句的状语，表示其所采用的方法，因此翻译时应将其前移，故整句应译为："运用诸如逐一支路算法、在少网孔配电网的潮流计算中的电压／电流迭代法，以及诸如回路分析法这样的直接法，首先计算节点注入的电流，再求解支路的电流，最后求解节点的电压，并以网络节点处的功率误差值作为收敛判据"。

【3】In the actual distribution system, some main lines have the automatic measuring table installation, but most of the distribution network can only be with the manual data collection, so it is difficult to ensure the accuracy of the run data, which limits the power flow calculation accuracy, making the majority of results can only serve as reference materials, and can not be used for actual decision-making.

该句很长，应仔细分清其框架结构。首先应该看出该句在主架结构上是一个由 but 连接的并列复合句。只不过后一个子句 most of the distribution network can only be with the manual data collection, so it is difficult to ensure the accuracy of the run data, which limits the power flow calculation accuracy, making the majority of results can only serve as reference materials, and can not be used for actual decision-making 比较长，其本身又是一个主从复合句。其中由 so 引导的结果状语从句 it is difficult to ensure the accuracy of the run data, which limits the power flow calculation accuracy, making the majority of results can only serve as reference materials, and can not be used for actual decision-making 是其主句 most of the distribution network can only be with the manual data collection 的状语。只不过该从句又是一个主从复合句，其中的 which limits the power flow calculation accuracy, making the majority of results can only serve as reference materials, and can not be used for actual decision-making 是其非限定性定语从句，该句的主语 which 有两个谓语动词 limites 和 can not be used，其中的 making the majority of results can only serve as reference materials 是第一个谓语动词的状语。分析清楚了以上情况后便可知，该句应译为："在实际的配电网中，只有部分主干线安装自动测量表计，而大部分配电网络只能通过人工采集网络的运行数据，因而很难保证其准确性。这就限制了配电网潮流计算结果的精确性，使得大多数计算结果只能作为参考资料，而不能用于实际决策"。

【4】As the power sections can only provide the load curve of distribution feeder root node and can not provide the load curves of each distribution transformer, based on the actual characteristics of distribution network, this paper distributes the total load at the root node according to a certain load distribution coefficient and the rated capacities of all transformers, and the obtained result has a smaller error compared with the actual situation.

该句可谓长句，翻译时首先应看清楚整个句子的结构。从整体上看该句为一主从复合句，其中 As the power sections can only provide the load curve of distribution feeder root node and can not provide the load curves of each distribution transformer 为原因状语从句，而 based on the actual characteristics of distribution network, this paper distributes the total load at the root node according to a certain load distribution coefficient and the rated capacities of all transformers, and the obtained result has a smaller error compared with the actual situation 为本句的主句。只不过该主句本身又很长，其中 based on the actual characteristics of distribution network 为一过去分词短语做其状语，this paper distributes the total load at the root node according to a certain load distribution coefficient and the rated capacities of all transformers 和 the obtained result has a smaller error compared with the actual situation 为其两个并列的子句。照此分析，可见该句应译为："由于供电部门只能提供配电馈线根节点处的总负荷曲线，不能提供每台配电变压器的负荷曲线，因此在计算时，本文根据实际配电网络的特点，将根节点处的总负荷按照一定的负荷分配系数，以及各配电变压器的额定容量进行分配，这样得出的计算结果，与实际情况相比误差较小"。

【5】But this way has a shortcoming that it does not take full advantage of a small number known load values in at some nodes and if with the state estimation method to estimate the node loads, the advantage of the known node load can be full taken and the calculation accuracy can be improved.

该句不是很长，也不是很难翻译，但有几处应值得注意。一是要看到句中的 that it does not take full advantage of a small number known load values in at some nodes 为 a shortcoming 的同位语从句。二是两个先后出现的并列连词 and 的翻译方法要注意，第一个 and 所连接的前后两个子句的关系是转折的，因而应翻译为"而"。后一个 and 所连接的前后两个子句的关系是因果的，因而应翻译为"从而"。故整句应译为："但这种方法的缺点是没有充分利用少数已知的节点负荷值。若运用状态估测法来估算各负荷节点的负荷，便可充分利用已知的节点负荷数据，从而可提高计算的精度"。

3.5.4　Translation

系统中的潮流计算

配电网潮流计算方法概况

目前，传统的电力系统潮流计算方法，如牛顿-拉夫逊法、PQ 分解法等，均以高压电网为其运用对象。而配电网的电压等级较低，其线路特性和负荷特性都与高压电网有很大区别，因此很难直接应用传统的电力系统潮流计算方法。由于缺乏行之有效的计算机算法，长期以来供电部门计算配电网潮流分布时大多数均采用人工手算方法。20 世纪 80 年代初期以来，国内外专家学者在手算方法的基础上提出了多种配电网潮流的计算机算法。目前辐射式配电网络潮流计算方法主要倾向于以下两类。

（1）直接应用基尔霍夫电压和电流定律

运用诸如逐一支路算法、在少网孔配电网的潮流计算中的电压／电流迭代法，以及诸如回路分析法这样的直接法，首先计算节点注入的电流，再求解支路的电流，最后求解节点的电压，并以网络节点处的功率误差值作为收敛判据。

（2）以有功功率 P、无功功率 Q 和节点电压平方作为系统的状态变量，列写出系统的状态方程，并采用牛顿-拉夫逊法求解该状态方程，即可直接求出系统的潮流解。

配电网络潮流计算的难点

数据采集

在配电网络潮流计算中，网络数据和运行数据的完整性和精确性是影响计算准确性的一个主要因素。对实际运行部门来说，要提供出完整和精确的配电网络数据和运行数据是很难办到的，这主要有下面几个原因。

（1）由于配电网的网络结构复杂，特别是10kV及以下电压等级的配电网络，用户多且分散，不可能在每一条配电馈线及分支线上安装测量表计，使得运行部门很难提供完整和精确的运行数据。

（2）在实际的配电网中，只有部分主干线安装自动测量表计，而大部分配电网络只能通过人工采集网络的运行数据，因而很难保证其准确性。这就限制了配电网潮流计算结果的精确性，使得大多数计算结果只能作为参考资料，而不能用于实际决策。

负荷的再分配

由于配电网络的网络结构复杂、用户设备种类繁多且极其分散、各种测量表计安装不全等原因，使得运行部门无法统计出每台配电变压器的负荷曲线，只能提供较为准确的配电网络根节点上（即降压变压器低压侧母线出口处）的总负荷曲线。因此在进行配电网络潮流计算时，采取何种负荷分配方法，尽可能真实地将根节点上的总负荷分配到各负荷节点上去，是潮流计算的关键问题。

实用的潮流计算程序

在配电网络潮流计算中，当三相配电系统负荷平衡时，其计算模型可由等值单相系统之模型来替代。其中，配电线用单位长度的电阻和电抗来表示，线路上的并联电容组的电容可暂时忽略不计（并联电容器组视为一负载）。对较长的辐射线路，线路容抗的注入量可认为是一电容负载。所有负载均不考虑其电压特性，即认为负载的功率不随电压幅值的变化而变化。

在对实际的配电网络进行潮流计算时，根据配电网络的实际和供电部门所提供的数据特点，在此仅做以下几点假设：

（1）假设配电网络为三相平衡网络，可用等值的单相网络来计算。

（2）假设所有的载荷配电变压器，在同一时刻为相同的负荷率。

（3）假定主线杆矩和分支线杆矩分别是固定的，这样可根据杆矩和总杆数计算出每条线路的长度。

（4）假设各负荷的功率因数相同。

（5）假设所有配电变压器根据各自的额定容量，在同一负荷率下，来分配根节点上的总负荷。

根据上述假设条件进行潮流计算时，需解决两个方面的实际问题：如何进行有效、正确、快速的支路追踪和进行负荷分配。由于配电网络是辐射型的，而且各配电馈线上分支线多、结构复杂，因此在计算时将每一条配电馈线的主馈线和各分支线均视为一条独立支路，分别计算其潮流值。计算时假设主馈线根节点（降压变压器母线出口处）的电压值恒定，各分支根节点电压（即主馈线上该节点电压）是已知的，且在分支潮流计算时不变。每次潮流计算时，先计算主馈线的潮流，而后更新各节点电压值（除主馈线根节点外），最后再计算所有各分支的潮流。

由于供电部门只能提供配电馈线根节点处的总负荷曲线，不能提供每台配电变压器的负荷曲线，因此在计算时，本文根据实际配电网络的特点，将根节点处的总负荷按照一定的负荷分配系数，以及各配电变压器的额定容量进行分配，这样得出的计算结果，与实际情况相比误差较小。但这种方法的缺点是没有充分利用少数已知的节点负荷值。若运用状态估测法来估算各负荷节点的负荷，便可充分利用已知的节点负荷数据，从而可以提高计算的精度。

在进行逆向计算时，若该分支根节点上有多条分支，则该分支根节点上的电压值，就取与其相连的所有分支所计算出来的根节点电压的平均值。而在进行正向计算时，则将根节点上的总功率按各分支负荷的分配系数分配到各分支上去。

根据上述思想，本文编制了计算程序，并以此对实际的配电网络进行了潮流计算，其计算结果符合实际情况，从而证明了上述方法的可行性。

数字仿真

根据上述分析，作者用 DistFlow 法、逐支路算法和牛顿-拉夫逊法分别对 IEEE 标准算例中的 69 个节点和 33 个节点的两个网络进行了潮流计算。表 1 是分别用这三种方法计算 69 个节点的网络的结果比较，表 2 是分别用这三种方法计算 33 个节点的网络的结果比较，表 3、表 4、表 5 则是采用 DistFlow 法和逐支路算法，对贵阳市北供电局管辖下的两条出线进行计算的结果分析。

从上述表中可看出，按照本文所采用的负荷分配方案，在支路参数 r/x 比值较大的情况下，由于配电网络本身呈现出病态网络的特征，采用标准牛顿-拉夫逊法计算时常常会出现发散或振荡的现象。而采用 DistFlow 法和逐支路算法计算就能较好地收敛，计算速度也快，收敛精度也高。但逐支路算法对某些网络（如 69 个节点网络）比较敏感，在收敛精度较高的情况下，会出现最大误差维持在一恒定值的现象，使得计算过程不能收敛。而采用 DistFlow 法不会出现这一现象。一般情况下，在相同的收敛精度下，逐支路算法比 DistFlow 法收敛速度要快。

Part 4 Electric Generation with Renewable Energy

Unit 4.1   Global Status of Wind Energy Market

Unit 4.2   Development on Electric Generation with Nuclear Power

Unit 4.3   Ecological Protection Project of Small Hydroelectric Power Replace of Fuel

Unit 4.4   Discussion on Biomass Power Generation

Unit 4.5   Hybrid PV-Battery-Diesel Power System

Unit 4.1　Global Status of Wind Energy Market

4.1.1　Text

One of the questions most often asked about wind power is "what happens when the wind does not blow". While on a local level this question is answered in chapter 5 (Grid integration), in the big picture wind is a vast untapped resource capable of supplying the world's electricity needs many times over. In practical terms, in an optimum, clean energy future, wind will be an important part of a mix of renewable energy technologies, playing a more dominant role in some regions than in others【1】. However, it is worthwhile to step back for a minute and consider the enormity of the resource【2】.

Researchers at Stanford University's Global Climate and Energy Project recently did an evaluation of the global potential of wind power, using five years of data from the US National Climatic Data Center and the Forecasts Systems Laboratory. They estimated that the world's wind resources can generate more than enough power to satisfy total global energy demand. After collecting measurements from 500 balloon-launch monitoring stations to determine global wind speeds at 80 meters above ground level, they found that nearly 13% had an average wind speed above 6.9 meters per second (Class 3), sufficient for economical wind power generation【3】.

North America was found to have the greatest wind power potential, although some of the strongest winds were observed in Northern Europe, while the southern tip of South America and the Australian island of Tasmania also recorded significant and sustained strong winds【4】. To be clear, however, there are extraordinarily large untapped wind resources on all continents, and in most countries; and while this study included some island observation points, it did not include offshore resources, which are enormous【5】.

For example, looking at the resource potential in the shallow waters on the continental shelf off the densely populated east coast of the US, from Massachusetts to North Carolina, the average potential resource was found to be approximately four times the total energy demand in what is one of the most urbanized, densely populated and highest-electricity consuming regions of the world【6】.

A study in German calculated that the global technical potential for energy production from both onshore and offshore wind installations was 278 000 TWh (Terawatt hours) per year【7】. The report then assumed that only 10%—15% of this potential would be realizable in a sustainable fashion, and arrived at a figure of approximately 39 000 TWh supply per year as the contribution from wind energy in the long term, which is more than double current global electricity demand【8】.

In summary, wind power is a practically unlimited, clean and emissions free power source, of which only a tiny fraction is currently being exploited.

The regions with the best wind regimes are located mainly along the southeast coast and the north and west of the country. Key provinces include Inner Mongolia, Xinjiang, Gansu Province's Hexi Corridor, some parts of North-East China, and the Qinghai-Tibetan Plateau.

Satisfying rocketing electricity demand and reducing air pollution are the main driving forces behind the development of wind energy in China. Given the country's substantial coal resources and the still relatively low cost of coal-fired generation, cost reduction of wind power is an equally crucial issue. This is being addressed through the development of large scale projects and boosting local manufacture of turbines.

The Chinese government believes that the localization of wind turbine manufacture brings benefits to the local economy and helps keep costs down. Moreover, since most good wind sites are located in remote and poorer rural areas, wind farm construction benefits the local economy through the annual income tax paid to county government.

The wind manufacturing industry in China is booming. In the past, imported wind turbines dominated the market, but this is changing rapidly as the growing market and clear policy direction have encouraged domestic production.

4.1.2　Specialized English Words

　　　　clean energy　清洁能源
　　　　global potential of wind power　全球潜在的风能储量
　　　　balloon-launch monitoring stations　气球高空监测台站
　　　　emissions free　无（核）放射
　　　　offshore resources　离岸海洋中的资源
　　　　shallow waters　浅水区域
　　　　sustainable fashion　可持续的方式
　　　　localization of wind turbine manufacture　风机制造的国产化
　　　　renewable energy　可再生能源
　　　　total global energy demand　全球能源需求总量
　　　　global wind speed　地球表面的风速
　　　　economical wind power generation　经济有效的风能发电
　　　　total installed capacity　总装机容量
　　　　continental shelf　大陆架
　　　　onshore and offshore　离岸海洋
　　　　ranks fifth　排列第五

4.1.3　Notes

【1】In practical terms, in an optimum, clean energy future, wind will be an important part of a mix of renewable energy technologies, playing a more dominant role in some regions than in others.

　　句中 in an optimum 意为"按照乐观的观点来看"，此处译为"乐观地预计"。a mix of renewable energy technologies，意为"可再生能源技术的综合体"，此处译为"整个可再生能源技术"。因此，整句译为："就实际情况而言，可乐观地预计，风能这种未来的清洁能源，将成为整个可再生能源技术中一个重要组成部分,其在某些领域中,将比其他形式的清洁能源发挥更重要的作用"。

【2】However, it is worthwhile to step back for a minute and consider the enormity of the resource.

句中 to step back 意为"走回来",此处译为"回过头来"。for a minute 意为"用几分钟",此处译为"花一点时间"。consider 意为"考虑",此处译为"回顾"。因此,整句译为:"无论怎么说,花一点时间回过头来回顾一下这种巨大的能源是值得的"。

【3】After collecting measurements from 500 balloon-launch monitoring stations to determine global wind speeds at 80 meters above ground level, they found that nearly 13% had an average wind speed above 6.9 meters per second（Class 3）, sufficient for economical wind power generation.

句中 after collecting measurements from 500 balloon-launch monitoring stations to determine global wind speeds at 80 meters above ground level, 意为"经过收集 500 个气球高空监测台站对距地面 80m 高空的地球表面风速的监测数据",此处译为"经过对来自 500 个气球高空监测台站,其对距地面 80m 高空的地球表面风速的监测所收集到的数据的分析"。此外,句中 sufficient for economical wind power generation 为形容词短语做全句的状语,对前述做补充说明。因此,整句译为:"经过对来自 500 个气球高空监测台站,其对距地面 80m 高空的地球表面风速的监测所收集到的数据的分析,他们发现其近 13%的平均风速超过了 6.9m/s（3 级）,这个风速足以满足经济有效的风能发电"。

【4】North America was found to have the greatest wind power potential, although some of the strongest winds were observed in Northern Europe, while the southern tip of South America and the Australian island of Tasmania also recorded significant and sustained strong winds.

该句为一主从复合句,主句为 North America was found to have the greatest wind power potential,而（although）some of the strongest winds were observed in Northern Europe, while the southern tip of South America and the Australian island of Tasmania also recorded significant and sustained strong winds,为其让步状语从句。因此,根据英汉两种语言之间的差异,汉译时应先译从句后译主句。此外还应看到,该从句本身又是一个主从复合句,应将其整体译完后再译全句的主句。因此,整句译为:"虽然所发现的某些最强劲的风是在北欧,而南美洲的最南端和澳大利亚的塔斯马尼亚岛也有相当可观和持续强风的记录,但是北美仍然是所发现的最大的潜在风能的所在地"。

【5】To be clear, however, there are extraordinarily large untapped wind resources on all continents, and in most countries; and while this study included some island observation points, it did not include offshore resources, which are enormous.

句中 to be clear 意为"要清楚",此处译为"应该清楚"。此外,and while this study included some island observation points, it did not include offshore resources, which are enormous 为与前句并列的子句,只不过该并列复合句中的并列子句本身又是一个主从复合句,while this study included some island observation points 为从句,it did not include offshore resources, which are enormous 为主句,而其中的 which are enormous 又为主句中的定语从句。明确了以上语法关系后,该句就不难翻译了。整句译为:"然而还应清楚:在各大洲的大多数国家,还蕴藏着巨大的尚未开发的风力资源。而这一研究仅仅是根据某些岛屿观察站的观察数据得出的结论,并不包括远离海岸线的大洋中的风能资源,而这将又是巨大的"。

【6】For example, looking at the resource potential in the shallow waters on the continental shelf off the densely populated east coast of the US, from Massachusetts to North Carolina, the average potential resource was found to be approximately four times the total energy demand in what is one of the most urbanized, densely populated and highest-electricity consuming regions of the world.

该句为一难长句,其中 looking at the resource potential in the shallow waters on the continental shelf off the densely populated east coast of the US, from Massachusetts to North Carolina 为一个现在分词短语,其做句子的状语。该句的句干为 the average potential resource was found to be。只是原句为被动语态,汉译时应按汉语的习惯转译为主动态,故整句译为:"例如,看看从马萨诸塞州到北卡罗来纳州,这一片人口稠密的美国东海岸大陆架上的浅水区域的潜在风力资源,就可发现其平均潜在资源几乎是一个在世界上城市化程度最高、人口最稠密、电力消费最高的地区所需能源总量的4倍"。

【7】A study in German calculated that the global technical potential for energy production from both onshore and offshore wind installations was 278 000 TWh(Terawatt hours)per year.

该句的句干为 A study in German calculated that,汉语中没有这样的表述方法,应给予转译,故整句译为:"据德国的一项研究推算,全球每年由沿海和离岸海洋中的风能设施所生产的具有技术性运用潜力的能源多达278 000TWh(万亿瓦小时)"。

【8】The report then assumed that only 10%-15% of this potential would be realizable in a sustainable fashion, and arrived at a figure of approximately 39 000 TWh supply per year as the contribution from wind energy in the long term, which is more than double current global electricity demand.

该句的主句 The report then assumed that only 10%-15% of this potential would be realizable in a sustainable fashion, and arrived at a figure of approximately 39 000 TWh supply per year as the contribution from wind energy in the long term 采用的是虚拟语态,是对将来可能发生的事情的一种假定,其中 would be 有两个表语 realizable 和 arrived,其为同位语关系。该句的从句 which is more than double current global electricity demand 则用的是现实语态,表示一种可实现的结果。其中 more than double 意为"超过两倍",故应译为"两倍多"。因此,全句译为:"该报告认为,只要这种潜在能源中的10%~15%以一种可持续的方式提供能源,也就是说只要在一段较长的时期内,每年(平均)由风能所提供的电力达到约39 000TWh(万亿瓦小时)这个数字的话,那么这些电能就已经是目前全世界电力需求的两倍多了"。

4.1.4 Translation

全球风能市场状况

关于风能,人们最常见的一个问题是"当风停止时将会是什么样的一种情形"。从小范围来说,这一问题已在第5章(电网的构成)中得到解决。而从大的范围来说,风是一种足以提供数倍于全球电能需求的巨大而尚未开发的能源。就实际情况而言,可乐观地预计,风能这种未来的清洁能源,将成为整个可再生能源技术中一个重要的组成部分,其在某些领域中,将比其他形式的清洁能源发挥更重要的作用。无论怎么说，花一点时间回过头来回顾一下这种巨大的能源是值得的。

最近，美国斯坦福大学全球气候与能源项目的研究人员，运用来自美国国家气候数据中心和预测系统实验室5年来的数据，做了一项全球潜在的风能储量的研究。他们估计，全球的风力资源对满足全球能源需求总量来说是绰绰有余的。经过对来自500个气球高空监测台站，其对距地面80m高空的地球表面风速的监测所收集到的数据分析，他们发现其中近13%的平均风速超过了6.9m/s（3级），这个风速足以满足经济有效的风能发电。

虽然所发现的某些最强劲的风是在北欧，而南美洲的最南端和澳大利亚的塔斯马尼亚岛也有相当可观和持续强风的记录，但是北美仍然是所发现的最大的潜在风能所在地。然而还应清楚：在各大洲的大多数国家，还蕴藏着巨大的尚未开发的风力资源。而这一研究仅仅是根据某些岛屿观察站的观察数据得出的结论，并不包括远离海岸线的大洋中的风能资源，而这将又是巨大的。

例如，看看从马萨诸塞州到北卡罗来纳州，这一片人口稠密的美国东海岸大陆架上的浅水区域的潜在风力资源，就可发现其平均潜在资源几乎是一个在世界上城市化程度最高、人口最稠密、电力消费最高的地区所需能源总量的4倍。

德国的一项研究推算了一下，全球每年由沿海和离岸大洋中的风能设施所生产的具有技术性运用潜力的能源多达278 000TWh。该报告认为，只要这种潜在能源中的10%~15%以一种可持续的方式提供能源，也就是说只要在一段较长的时期内，每年（平均）由风能所提供的电力达到约39 000TWh这个数字的话，那么这些电能就已经是目前全世界电力需求的两倍多了。

总而言之，风力发电是一个取之不尽的、清洁无污染、无（核）放射的能源，然而目前只利用了其中的很小一部分。

中国的最佳风力资源地区主要位于东南部沿海海岸，以及大陆的北部和西部。主要省份包括内蒙古自治区、新疆维吾尔自治区和甘肃省的河西走廊、中国的东北一些部分，以及青藏高原。

满足迅速发展的电力需求和减少空气污染是中国风能发展的动力。降低风力成本与给国家提供大量煤炭资源，进而提供成本相对较低的燃煤发电是同样重要的问题。目前，这一目标正在通过推广大型项目和促进国内风机生产的两项措施加紧实现。

中国政府相信，风机制造的国产化会给国内经济带来好处，同时也有助于降低风力发电成本。此外，由于大多数好的风场位于偏远和贫困的农村地区，因此风力发电场的建设将通过每年交给政府的所得税，对当地经济作出贡献。

风力制造业在中国蓬勃发展。过去，进口的风力涡轮机在市场占主导地位，但随着市场的不断扩大，这一状况正在发生迅速变化，国家明确的政策导向也是鼓励国内生产。

Unit 4.2　Development on Electric Generation with Nuclear Power

4.2.1　Text

By 2020, with the economic development objectives, the quadrupling China GDP is estimated that China power generation capacity need is from the current 400GW to 2020 960GW about the need for new 560GW.

China is meeting three major challenges facing in the energy supply. The first one is a very acute contradiction between energy supply and demand, the second, the contradiction between the demand for energy and the rather less resource reserves per capita. And the third one is the coal-dominated energy structure is irrational, which causes a large number of serious environmental pollution a serious greenhouse gases problem with the coal-firing. To cope with these challenges, China must take active measures to gradually change the current energy structure unreasonable.

In China, the current power supply structure is that in which the coal-firing electricity power supply accounts for 74%, while hydropower 24% and nuclear power only 1.6%. With the development on the hydroelectric power, nuclear energy, the non-water renewable energy (especially wind) and other initiatives, we can gradually reduce the share of fossil fuels and improve the energy structure. Taking into account China energy structure situation history and reality, by 2020, China energy supply will be in the pattern of coal-based. That is to say in the new increasing 560GW, more than half of it will remain dependent on coal. By 2020, even if the new installed capacity of the hydropower electric generation is about 160GW, there is still a large gap in the demand for electricity supply. The gap must be filled by nuclear power, that is, by 2020, China nuclear power installed capacity should reach about 40 million kW (about 4% in the national installed electric generators). This requires, during the next 10 years, about 30 new million nuclear power plants to be built, and of which, from now, 2-3 million kW nuclear power plants have to start building, the task being very difficult.

Taking into account the national security, energy security and environmental security issues, and taking into account after 2020, the oil and gas will become more dependent on the international market, the coal-fired power generation on the environment will be greater pressure and the development of hydropower will also be without too much room for, as a large-scale development of alternative energy sources, it is unquestionable that the nuclear energy will certainly be in an important position in China energy development 【1】.Huge market demand for nuclear power in China's nuclear power industry based on the survival and development, but also to China's nuclear power industry a comprehensive, coordinated and sustainable development put forward higher requirements.

At present, China nuclear power accounts for 1.6% of its total generating capacity, and the nuclear power proportions of both Zhejiang and Guangdong provinces all have reached 13%, or nearly the world average 16 percent. The operations of all nuclear power plants in China are good, and some of their performances have reached the advanced international level. The independent constructed Qinshan Nuclear Power Station Phase I has been safely operated for 13 years. The Phase II and Phase III also have fully completed and put into production ahead of schedule, which is a great forward step in China commercial nuclear power plant self-construction and a new record in the international nuclear power plant constructions of same types. Besides this, Daya Bay Nuclear Power Station have been operating safely and stably for ten years and Ling Ao Nuclear Power Station has been fully completed and put into production, whose performances are in the top of the international similar power plant operation.

China nuclear power industry development history is nearly 20 years. With the implementation of nuclear power projects and the key technology study, in the nuclear power technology developing, engineering design, equipment manufacturing, construction, project management, nuclear power plant operation and maintenance, and so on, China has a good base and a strong power, with which China has 300 000 kW nuclear power plant completed sets export capacity and 600 000 kW nuclear power plant own capacity building and design capacity. Of course, compared with developed countries, we still have a gap in the nuclear power design and manufacture level.

With fully utilizing the opportunities of reforming and opening up, we are actively developing the new international cooperation situation to shorten, as soon as possible, the gap between the world level and ours. With the international cooperation and the introduction of their technology, we can put the advance large-scale pressurized water reactor nuclear power technology (meet with the level of the third generation technology) industrialization starting point onto a new higher platform, and have, as soon as possible, the capacity in own-design, own-manufactured, own-building and own-run of the large-scale advanced pressurized water reactor nuclear power plant, which then gradually form a nuclear power technology with our own brand and intellectual property rights. We can not wait for and rely on the others. By 2020, we want have the building, in a mass, ability of the nuclear power plants meeting the third technical requirements.

Fast reactor, as a follow-up to the development of advanced pressurized water reactor type, enters the business applications in about 2035 is appropriate. For this, we plan to build a prototype fast reactor by 2020, so that, after 2040, the fast reactor nuclear power system can be produced in mass, and after 2050, they can gradually become the main force of China nuclear energy. Closed circulatory system the nuclear fuel (including the high-level radioactive waste treatment and disposal) must be simultaneously coordinated development.

We believe that only with such a deployment, China nuclear power industry can meet the comprehensive and coordinated scientific development requirement. Then we can not only catch up with international advanced level in time, but also have adequate reserve strength in the sustainable development.

4.2.2 Specialized English Words

alternative energy sources　可替代能源
fast reactor　快堆（核电技术）
radioactive waste treatment and disposal　高放射性废物处置
pressurized water reactor　压水堆（核电技术）
closed circulatory system　闭合循环系统

4.2.3 Notes

【1】Taking into account the national security, energy security and environmental security issues, and taking into account after 2020, the oil and gas will become more dependent on the international market, the coal-fired power generation on the environment will be greater pressure and the development of hydropower will also be without too much room for, as a large-scale development of alternative energy sources, it is unquestionable that the nuclear energy will certainly be in an important position in China energy development.

该句为一难长句，整句的句干为 it is unquestionable，其中的 it 是先行词，替代由 that 引出的主语从句 that the nuclear energy will certainly be in an important position in China energy development。而整个句子的原因状语为均由 taking 引出的两个并列的分词短语，只不过第一个分词短语中的宾语是三个并列的名词性短语 the national security, energy security and environmental security issues，而第二个分词短语中的宾语则为三个并列的宾语从句 after 2020, the oil and gas will become more dependent on the international market, the coal-fired power generation on the environment will be greater pressure and the development of hydropower will also be without too much room for。因此，全句应译为："考虑到国家安全、能源安全、环境安全等问题，到 2020 年以后石油和天然气将更多地依赖国际市场，燃煤发电对环境的压力将更大，水能的开发也将无太多余地，因此作为可大规模发展可替代能源的计划，毋庸置疑，核能必将在国家能源发展中占据重要的地位"。

4.2.4 Translation

大力发展核能发电

到 2020 年，按中国 GDP 经济发展翻两番的目标来估计，我国的发电装机容量需从现在的 400GW 提高到 2020 年的 960GW 左右，即需新增 560GW。

中国的能源供应面临着三大挑战：第一，能源供需矛盾极为尖锐；第二，能源发展需求与中国能源资源储量的人均拥有量相对不足之间也存在着尖锐的矛盾；第三，以煤炭为主的能源结构不合理，大量燃煤造成严重环境污染和严重的温室气体效应问题。为应对上述挑战，中国必须采取积极措施，逐步改变目前不合理的能源结构状况。

目前，在中国的电力供应中，煤电占 74%，水电占 24%，核电仅占 1.6%。通过大力发展水电、加快发展核电、积极发展非水可再生能源（尤其是风能）等举措，可以逐步降低化石燃料的份额，逐步改善能源结构。考虑到中国能源结构的历史与现实状况，2020年之前中国能源供应仍将无法摆脱以煤炭为主的格局。即在新增加的560GW中，将有一半以上仍依赖于煤电。到2020年，即使水电装机容量新增160GW左右，仍然存在很大的电力需求缺口。这个缺口必须由核电来填补，即到2020年，中国核电装机容量应达到40 000MW左右（届时约占全国总装机容量的4%）。这就需要在今后的10年间，新开工建设30台左右的百万级核电机组，也就是从现在起，每年要开工建设2~3台百万千瓦级的核电站，任务非常艰巨。

考虑到国家安全、能源安全、环境安全等问题，考虑到2020年以后石油和天然气将更多地依赖国际市场，燃煤发电对环境的压力将更大，水能的开发也将无太多余地，因此作为可大规模发展可替代能源的计划，毋庸置疑，核能必将在国家能源发展中占据重要的地位。庞大的核电市场需求是中国核能工业生存和发展的基础，同时也对其全面、协调和可持续发展提出了更高的要求。

目前，中国核发电量占总发电量的1.6%，浙江和广东两省的核电比例达到13%，已接近世界平均16%的水平。中国所有核电站的运行业绩良好，部分运行指标已达到国际先进水平。第一座自主建设的秦山一期核电站已经安全运行13年。秦山二期核电站也全面建成投产，这是中国自主建设的商用核电站向前跨出的重大步骤。秦山三期重水堆核电站提前建成投产，创造了国际同类型核电站建设的多项纪录。大亚湾核电站保持10年安全稳定运行，岭澳核电站也已经全面建成投产，其运行业绩在国际同类电站中名列前茅。

中国发展核电工业已有近20年的历史，通过已建和在建核电项目的实施及关键技术的研究，中国在核电技术开发、工程设计、设备制造、工程建设、项目管理、核电站运行维修等方面已经具有较好的基础和较强的实力，已经具备30万千瓦核电站的成套出口能力和60万千瓦核电站的自主设计建设能力。当然，与发达国家相比，我们在核电设计和制造水平方面还存在一定差距。

我们正在充分利用改革开放的机遇，积极开拓国际合作新局面。尽快缩短与世界水平的差距。通过国际合作，引进技术，我们可以将大型先进压水堆核电技术（符合国际上第三代技术水平）的自主化和产业化的起点构筑在新的更高的平台基础上，并尽快具备自主设计、自主制造、自主建设、自主运行大型先进压水堆核电站的能力，逐步形成具有自主知识产权的核电品牌。我们一不等，二不靠。到2020年，中国要具备批量建设符合国际上第三代技术要求的核电站的能力。

快堆作为先进压水堆的后续发展堆型，在2035年前后进入商业应用是适宜的。为此，计划在2020年左右建成原型快堆，使快堆核能系统在2040年以后得以批量发展，并在2050年以后逐步成为中国核能主力。相应的核燃料闭合循环系统（包括高放废物处置）也必须同步协调发展。

我们认为，只有这样的部署，中国的核电工业才能符合全面和协调发展的要求。这样，我们不仅可以适时赶上国际先进水平，而且将具备可持续发展的充足的后劲。

Unit 4.3 Ecological Protection Project of Small Hydroelectric Power Replace of Fuel

4.3.1 Text

The project of small hydropower replacing firewood for ecology protection (SHPF) is a public interests project that consolidate the achievements of returning land for farming to forestry and solve the problem of the fuel for people life. At present, the experimental unit work of SHPF project is just start, and the study on evaluation of ecology environmental benefits, determination of electricity price and comprehensive evaluation is little. So it has very important academic value and practical meaning .This thesis use the technical route that combine the theoretic study and case analysis, based on the material of home and abroad, use the cross theory of the resources and environmental economies, ecology economies, ecology, hydrology and water resources, waterpower utilization, rural energy and system engineering, and take the Magedang in Shanxi Province as an example to make a deep study in ecology benefits, electricity pricing, comprehensive evaluation and system dynamics simulation 【1】. The main achievements are as follow:

(1) Power demand without SHPF is forecasted by Grey system:Taking the measure of combining stratified sampling and typical survey to make a survey analysis of the energy structure in Magedang, the results show that before the SHPF Project carry out, the main fuel of peasant life in experimental area is firewood, and the firewood mainly form the natural reserve, state owned forest, public forest. After carry out the SHPF Project, the replacing fuel price accordance with the peasant afford ability and people life energy gradually developing into use the electricity as main fuel. Because the most SHPF projects are run-of-rive power station, short of electricity in dry season, so we should adjust measures to local conditions and found energy complement system.

(2) This thesis studies the method of the economic increasing effect of SHPF Project. With the directory analysis method, from the expenses of the SHPF construction and the input of the labor, construction material and machine decrement, the economic increasing effect of construction SHPF to national economy is studied. With the region investment multiplier method, the economic increasing effect to GDP in Magedang in Shanxi Province is analysed. The results show that as create job opportunity and increasing GDP, the construction of SHPF not only developing the natural resources, but also increasing the opportunity of local peasant to work and promoting the development of economy of experimental area.

(3) This thesis studies the method of ecology benefits evaluation found model based on promote people healthy life level, reduce carbon dioxide and sulfur dioxide .In this method, we divide the benefits into a lot of parties: water-holding, soil and water conservation oxygen-making, carbon dioxide and sulfur dioxide absorbing and atmosphere purification benefit.

(4) Based on analysis merits and demerits of the single electricity price and two-part electricity

price, this thesis establishes the three-part electricity price model that includes capacity, energy and ecology price, adopts combination method of average cost and marginal cost to study the problem of electricity price, and with the average increasing benefits method, solves the problem which hard to obtain with the ecology marginal benefits 【2】.The model has been used to calculate electricity price in Magedang experimental area.

(5) With the quantification research of the ecology and social benefits, the thesis establishes the comprehensive evaluation system, which includes three layers (the effete of economic, ecology environmental benefits and social benefits), four Parts (the effete of economic cost, ecology environmental benefits, energy development effect, social benefits), and 24 indexes, puts forward the evaluation standard and method of each index and according to the layer characteristics of comprehensive evaluation system, establishes the fuzzy hierarchy analysis model and evaluates two small hydropower stations in experimental area 【3】.

(6) With system dynamics method, this thesis studies the simulation of SHPF development, in which we establish the consequence loop of the project, system dynamics flow diagram model, the determined method of the parameters The model applied in Magedang combined with the Shanxi practical situation. The result shows that system dynamics theory simulation is an effective method to study the SHPF project scheme evaluation.

The development, the energy, the environmental protection are the current Chinese most popular topic. To promote the national economy good and quick developing, the relation within these three aspects needs to deal with coordinately very well. With the national economy keeping quite and high growth, the electric power demand will rapidly growth. The hydraulic electro generating is clean, does not have the pollution, may circulate the use, costs lowly, is high benefit and has small effect to the environmental. For the small hydroelectric power station, because its development scale is smaller than the large one, its influence onto the environment is also smaller than the latter. Therefore the product of the small water power station is one kind of green energy.

The so-called hydraulic electro generating is (has flood peak) using the water power to impel the water frame (hydraulic turbine) to rotate, with which the hydro energy can transform into the mechanical energy, while another kind of machinery (generator) on the hydraulic turbine axle then transforms it into electricity, with its rotating with the hydraulic turbine 【4】.That is to say the hydraulic electro generating is a process in which the water potential energy turns into the mechanical energy and then turns into the electrical energy.

The small electricity generation with water power belongs to the non-carbon clean energy. Since it not only has not the problem in resource drying up, but also has no pollution to the environment, The small hydraulic electro generating is the essential constituent in China implementing the sustainable developmental strategy. It can use the resources reasonably and promote the society, economy and environment coordinately and sustainably developing, in a good ecological environment foundation. It is important not only to escaping from poverty, become rich and raising the people live level for the countryside area, but also to protecting the ecological environment, and to promoting the coordinately development between the countryside society and the economic environment, to develop the small electro generating with water power, the renewable energy, under the circumstance permit and transform the hydro resources to the high grade electrical energy.

4.3.2 Specialized English Words

Small Hydropower Replacing Firewood （SHPF）替代燃烧柴草的小水电工程
ecology protection 生态保护工程
returning land for farming to forestry 退耕还林
evaluation of ecology environ mental benefit 生态环境效益评估
technical route 技术路线
ecology economies 生态经济学
ecology 生态学
waterpower utilization 水能利用
system engineering 系统工程
run-of-rive power station 径流式水电站
directory analysis method 直接分析法
create job opportunity 创造就业机会
carbon dioxide 二氧化碳
found model 建模
single electricity price 一部制电价
three-part electricity price model 三重结构的电价模型
energy price 电量电价
average cost 会计（平均）成本
average increasing benefits 平均增量效益
comprehensive evaluation system 综合评价指标体系
have flood peak 具有水头
hydraulic turbine 水轮机
firewood 柴草
small hydropower 小水电
public interests project 公益性工程
fuel for people life 生活用燃料
experimental unit work 试点工作
comprehensive evaluation 综合评价
material of home and abroad 国内外的文献资料
resources and environmental economies 资源环境经济学
hydrology and water resources 水文水资源
rural energy 农村能源
Grey system 灰色系统理论
energy complement system 多能互补体系
region investment multiplier Method 区域投资乘数法

increasing GDP GDP 增长
sulfur dioxide 二氧化硫
two-part electricity price 二部制电价
capacity price 容量电价
ecology price 生态电价
marginal cost 边际成本
ecology marginal benefits 生态边际效益
hydraulic electro generating 水力发电
water power 水能
water frame 水力机械

4.3.3 Notes

【1】This thesis use the technical route that combine the theoretic study and case analysis, based on the material of home and abroad, use the cross theory of the resources and environmental economies, ecology economies, ecology, hydrology and water resources, waterpower utilization, rural energy and system engineering, and take the Magedang in Shanxi Province as an example to make a deep study in ecology benefits, electricity pricing, comprehensive evaluation and system dynamics simulation.

本句为一难长句，其中 based on the material of home and abroad 为一过去分词短语，做全句的状语。主句的主语是 This thesis，其后依次连接三个并列的谓语动词 use..., use..., and take。只是在第一个并列子句中又有一个定语从句 that combine the theoretic study and case analysis，其修饰第一个谓语动词 use 的宾语 the technical route。此外，第二个谓语动词 use 的宾语 the cross theory 所指的交叉学科的理论很多，其包括 the resources and environmental economies, ecology economies, ecology, hydrology and water resources, waterpower utilization, rural energy and system engineering, 因而显得句子很长。但若将以上分析清楚了，本句的翻译也是不难的。照此分析，本句可译为："本文在查阅大量国内外相关文献资料的基础上，采用理论研究和实例分析相结合的技术路线，运用资源环境经济学、生态经济学、生态学、水文水资源、水能利用、农村能源和系统工程等交叉学科的理论与方法，以山西省陵川县马圪当乡试点区为例，对小水电替代燃烧柴草工程生态环境效益评估方法、电费定价、综合评价指标体系的建立和评价方法，以及系统动力学仿真等方面进行了深入研究"。

【2】Based on analysis merits and demerits of the single electricity price and two-part electricity price, this thesis establishes the three-part electricity price model that includes capacity, energy and ecology price, adopts combination method of average cost and marginal cost to study the problem of electricity price, and with the average increasing benefits method, solves the problem which hard to obtain with the ecology marginal benefits.

本句也是一个难长句，其中 based on analysis merits and demerits of the single electricity price and two-part electricity price 为过去分词 based 引导的过去分词短语，做整句的条件状语。本句主句的主语 this thesis 有三个并列的动词做其谓语，这就是：establishes、adopts 和 solves。只是在第一个谓语动词的宾语 three-part electricity price model 后有一个定语从句 that includes capacity, energy and ecology price。而在第二个谓语动词后又有一个不定式短语 to study the problem of electricity price，其为该谓语动词的目的状语。此外，在第三个谓语动词的宾语 the problem 后，又有一个定语从句 which hard to obtain with the ecology marginal benefits。由于这一系列的原因，使得句子显得很长。但照此详尽分析后便不难翻译了，此句可译为："在分析一部制电价和二部制电价优缺点基础上，本文建立了由容量电价、电量电价和生态电价三重结构组成的三部制小水电替代燃烧柴草的电价模型，采用会计成本和边际成本相结合的方法，研究了小水电替代燃烧柴草工程电价问题，同时采用平均增量效益解决了生态边际效益难于获得的问题。"

【3】With the quantification research of the ecology and social benefits, the thesis establishes the comprehensive evaluation system, which includes three layers (the effete of economic, ecology environmental benefits and social benefits), four Parts (the effete of economic cost, ecology environmental benefits, energy development effect, social benefits), and 24 indexes, puts forward the evaluation standard and method of each index and according to the layer characteristics of comprehensive evaluation system, establishes the fuzzy hierarchy analysis model and evaluates two small hydropower stations in experimental area.

本句又是一个难长句，其中由介词 with 引导的介词短语 with the quantification research of the ecology and social benefits 为整句的条件状语。而主句的主语 the thesis 后有四个连续的动词 establishes、puts、establishes 和 evaluates 做其谓语。只不过在第一个谓语动词的宾语 comprehensive evaluation system 之后又有一个很长的非限定性定语从句 which includes three layers (the effete of economic, ecology environmental benefits and social benefits), four Parts (the effete of economic cost, ecology environmental benefits, energy development effect, social benefits), and 24 indexes。此外，在最后的两个谓语动词之前又有一个由现在分词 according 引导的现在分词短语 according to the layer characteristics of comprehensive evaluation system 做该两个动作的条件状语。综上所述，这么多的修饰与被修饰关系错综复杂，致使全句很长。但只要进行以上分析便不难发现，本句应译为："在研究该工程的生态效益和社会效益量化评估的基础上，本文建立了一个对该工程进行综合评价的指标体系。该系统包括三个层次（即经济效益、生态环境、社会效益）分四个部分（即经济投资效果、生态环境效益、能源开发效果和社会效益），以及24项指标。此外，还提出了各指标的评价标准和贯彻每项指标的方法。根据综合评价指标体系的层次结构特点，建立了模糊层次分析模型，并以此对试点区的两座替代燃烧柴草的小水电站进行了评价。"

【4】The so-called hydraulic electro generating is (has flood peak) using the water power to impel the water frame (hydraulic turbine) to rotate, with which the hydro energy can transform into the mechanical energy, while another kind of machinery (generator) on the hydraulic turbine axle then transforms it into electricity, with its rotating with the hydraulic turbine.

该句为一复合句，其主句为 The so-called hydraulic electro generating is (has flood peak) using the water power to impel the water frame (hydraulic turbine) to rotate。而由介词 with 引导的 with which the hydro energy can transform into the mechanical energy 为一介词短语，在句中做状语，表示一种伴随状态。只不过该短语中介词的宾语是一宾语从句 which the hydro energy can transform into the mechanical energy。此外，该句中还有一个状语，那就是由从属连词 while 引导的时间状语从句 while another kind of machinery (generator) on the hydraulic turbine axle

· 143 ·

then transforms it into electricity, with its rotating with the hydraulic turbine，也是表示此时的一种伴随状态。只不过该从句中还有自己的状语，这就是由with引导的介词短语with its rotating with the hydraulic turbine。只不过该介词短语中又有一个介词短语with the hydraulic turbine做自己的方式状语。如此分析起来，可见该句之复杂。但若以上关系清楚了，也就看出了其应译为："所谓水力发电就是利用水能（具有水头）推动水力机械（水轮机）转动，将水能转变为机械能，而与水轮机同轴相连另一种机械（发电机）便随水轮机转动，从而将机械能转变为电能"。

4.3.4　Translation

运用小水电替代柴草的生态保护工程

小水电替代燃烧柴草的生态保护工程是巩固退耕还林成果，解决农民生活用柴问题的公益性工程。目前，小水电替代燃烧柴草生态保护工程试点工作刚刚起步，而对小水电替代燃烧柴草工程的生态环境效益评估、电价的确定，以及综合评价等问题的研究甚少。因此，开展此项研究具有十分重要的学术价值和实际意义。本文采用理论研究和实例分析相结合的技术路线，在查阅大量国内外相关文献资料的基础上，运用资源环境经济学、生态经济学、生态学、水文水资源、水能利用、农村能源和系统工程等交叉学科的理论与方法，以山西省陵川县马圪当乡试点区为例，对小水电替代燃烧柴草工程生态环境效益评估方法、电费定价、综合评价指标体系的建立和评价方法，以及系统动力学仿真等方面进行了深入研究。论文的主要研究成果为：

（1）用灰色系统理论预测了尚未实施小水电替代燃烧柴草工程情况下的农村用电量情况。采用分层抽样和典型调查相结合的方法，对山西省马圪当乡项目区的能源结构进行了调查研究，结果表明：在实施小水电替代燃烧柴草工程以前，项目区的农民生活能源以柴草为主。而这些柴草又主要来自自然资源、国家森林以及集体林木。小水电替代柴草工程实施以后，替代燃烧柴草的电价与农民承受能力基本相符，农民的生活能源逐渐发展为以电力为主。因为替代柴草的水电站多为径流式水电站，其在枯水期供电不足，所以应该调整策略，因地制宜，建立多能互补体系。

（2）本文对小水电替代燃烧柴草生态保护工程对经济拉动作用的评估方法进行了探讨。采用直接分析法，从分析替代柴草水电站建设的开支、建设中的劳动力投入，以及机械设备的损耗等入手，研究了小水电替代柴草生态保护工程对国民经济的拉动作用。采用区域投资乘数法，分析了山西省马圪当乡小水电替代燃烧柴草料项目建设对国民经济的拉动作用。结果表明，就创造就业机会和拉动GDP增长而言，替代燃烧柴草的水电站建设不仅开发了当地丰富的自然资源，而且增加了当地农民的就业机会，带动了项目区经济的发展。

（3）本文探讨了生态效益的评估建模方法，并以此分析该项工程实施后对提高农民健康水平、减少二氧化碳和二氧化硫排放量所起的作用。在运用该方法的分析过程中，将保护森林的效益分为涵养水源、水土保持、制造氧气、吸收二氧化碳和二氧化硫，以及净化大气等。

（4）在分析一部制电价和二部制电价优缺点基础上，本文建立了由容量电价、电量电价和生态电价三重结构组成的小水电替代燃烧柴草电价模型，采用会计成本和边际成本相结合的方法，研究了小水电替代燃料工程电价问题，同时采用平均增量效益解决了生态边际效益难于获得的问题。该模型已用于马圪当乡小水电替代柴草试点区的电价计算。

（5）在研究该工程的生态效益和社会效益量化评估的基础上，本文建立了一个对该工程进行综合评价的指标体系。该系统包括三个层次（即经济效益、生态环境、社会效益）分四个部分（即经济投资效果、生态环境效益、能源开发效果和社会效益），以及 24 项指标。此外，还提出了各指标的评价标准和贯彻每项指标的方法。根据综合评价指标体系的层次结构特点，建立了模糊层次分析模型，并以此对试点区的两座替代燃烧柴草的小水电站进行了评价。

（6）采用系统动力学方法，本文对替代燃烧柴草的小水电生态保护工程的进展情况进行了仿真研究。在此期间，我们建立了该工程的链式结构的系统总图、系统动力学流程图模型，以及仿真模型参数的确定方法。该仿真模型已结合山西省的实际情况，对马圪当乡项目进行了仿真。其结果表明，系统动力学仿真理论是研究替代燃烧柴草的小水电工程建设项目评价的有效方法。

发展、能源、环保是当前中国最热门的话题。为了促进国民经济又好又快地发展，需要协调好以上三个方面的关系。随着国家经济的持续高速发展，对电力的需求必将进一步加大。水力发电清洁、无污染、可循环利用、成本低、效益高、对环境影响小。对于小型水电站，因为其开发规模小，因而对环境的影响相对于大型水电站来说要小得多，所以小型水力发电生产的是一种绿色能源。

所谓水力发电就是利用水能（具有水头）推动水力机械（水轮机）转动，将水能转变为机械能，而与水轮机同轴相连另一种机械（发电机）便随水轮机转动，从而将机械能转变为电能。也就是说，水力发电是将水的势能转变为机械能，而后又将其转变为电能的转换过程。

小型水力发电属于非碳清洁能源。既不存在资源枯竭问题，又不会对环境造成污染，是中国实施可持续发展战略不可缺少的组成部分。也是在良好的生态环境基础上，合理利用资源，促进社会、经济和环境协调与可持续发展。因地制宜地开发小水电这种可再生能源，将水力资源转变成高品质的电能，不仅对于农村地区的脱贫致富，提高人民生活水平，而且对保护生态环境，促进农村社会与经济环境的协调发展都有着十分重要的意义。

Unit 4.4　Discussion on Biomass Power Generation

4.4.1　Text

In the renewable energy power generation, the biomass power generation has the characteristics of better quality, higher reliability, and more mature technology, and it can be achieved without intermittent power generation. At present, many areas can not effectively make use of straw, which just be burned, resulting in flue gas pollution in a wide range. Therefore, being suited to local conditions, to make use of biomass for power generation and to transmission to the power grid or used as a distributed power supply, is a good way of turning the waste into the useful things and of saving energy, reducing environmental pollution, benefiting to the farmers【1】. The development of biomass power generation can promote the development of circular economy. And it is also useful to building a kind of resource-saving and environment-friendly society and to building a new socialist countryside. At present, most of countries in the world take the biomass as an important alternative energy sources.

1. The need and feasibility of the development in biomass power generation in our country

China has abundant biomass resources and the total biomass in theory is close to 1500 million tons of standard coal. By 2020, the biomass development and utilization amount will be up to 500 million tons of standard coal, which is equivalent to more than 15% increase in energy supply. Compared with the traditional fossil fuels, biomass energy is clean energy and the carbon dioxide emissions after the burning belong to the natural world carbon cycle, which will not formate the new pollution the natural world. The biomass sulfur content is very low, only 3%, less than a quarter of the one in the coal. The development of biomass power generation and the implementation of alternatives to coal can significantly reduce carbon dioxide and sulfur dioxide emissions, resulting in huge environmental benefits.

China has 1800 million MU (0.0667 hectare) of sowing area of crops, resulting in about 700 million tons of farm material, within which except a part of for the paper-making and livestock feed, the last one all for the fuel. On the other hand, China existing forest area is about 175 million hectares and its forest cover rat is 18.21%. In a year, the available biomass resources are about 800 million to 1000 million tons. All of these are the good base of the development in the electric generation with biomass power.

2. Biomass power generation in China

China attaches great importance to the bio-power generation. Under the guidance of the policy, China electric generation with bio-power industry has developed rapidly. In recent years, State Grid Corporation, the top five power generation groups and other large state-owned, the private and foreign enterprises have investment in biomass power generation to participate in China construction industry operators.

In the course of biomass power industry development, the most prominent is the State Bio-Power Company Limited under the State Grid Corporation. At the present, this company has approved 39 biomass power generation projects in more than 10 provinces in the country. Among them, 10 projects have been completed and put into production, with which these capacities are 2 500 million watts, and 9 projects is in the installing, with which these capacities are 108 million watts. In 2007, the consumption of agricultural and forestry waste is about more than 1.3 million tons and for this, the local farmer income increases nearly more than 300 million Yuan, the replaced standard coal is more than 0.6 million tons and the reduced carbon dioxide is more than 0.6 million tons. In Dan County, Shandong Province, a biomass power generation project that was equiped with 1×25 million kilowatts units was formally put into operation on December 1, 2006, creating the first national direct-fired biomass power generation and opening up a new way to benefit farmers. This project electricity generation capacity in design is 160 million kilowatts, while the electricity generation was up to 229 million kilowatts in 2007, and counting according to the installed capacity 25 million watts, the number of operation hours throughout the year was up to 9160 hours, which reached the world advanced level and created a miracle in electricity generation with biomass 【2】. The company investment project in Wangkui, Heilongjiang Province, is the first biomass power generation project in the world that uses corn stalk for the main fuel, which was formally completed and put into production on November 16, 2007, since that time, it has been running stably, and which opens up a new way of corn and the such yellow straws comprehensive utilization, for the Northeast and Central Plains region in China and has contributed to the development of direct-fired biomass for electric power generation industry for our country and the world 【3】.

3. Prospects for biomass power generation in China

Biomass is a traditional renewable energy sources. Before Industrial Revolution, it was the major energy source supporting the development of human society. At present, as a result of the emergence of fossil fuels, the amount of biomass energy used currently is very small. However, with the energy resources and energy environmental issues becoming increasingly prominent, the use of biomass energy have once again become a choice of the mankind. If more advanced technology will be used, the products can partly replace the existing electric power that from consuming the fossil fuels. Therefore, the use of biomass energy should be over-all planning and all-round considerations. In the biomass-rich regions, such as the major grain-producing areas, large-scale food processing enterprises, forest, large-scale wood processing plants and the like, a number of biomass power stations can be building according to the local conditions, while in most of the resources dispersed rural areas, with solving the fuel problem, the use of biomass forming particles fuel or biogas technology should be promoted, to solve the problem of decentralized fuel consuming with the distributed resources 【4】.

The Chinese Government has given a strong support to the biological power generation industries. It is worth note that, as a new type industry, to multi-joint and coordinate within the cross-course, cross-section and cross-profession and to reconcile the contradiction between the fuel demand and lagging production patterns in rural areas are the issues requiring urgent solution 【5】. If you overcome these difficulties, biomass power generation will be in its modernization drive to play a greater role.

4. Development in biomass power industry abroad

Since 1990, the biomass power generation has begun large-scale development in many countries in Europe and America. Especially since Johannesburg World Summit on Sustainable Development in 2002, the development and utilization of biomass energy has been accelerating in the world. By the end of 2004, the biomass power generation capacity in the world had reached 39 000 million watts, about 200 billion kWh in electricity generation in a year, which can substitute 70 million tons of standard coal and is the sum quantity of the generating electricity in the wind power, photovoltaic, geothermal and other renewable energy 【6】.

Finland is one of the best successful countries using biomass for power generation in EU, whose current biomass power generating capacity accounts for all the national capacity 11%. Austria successful implements a plan of establishing the regional power stations with burning wood residue, with which the biomass in the proportion of total energy consumption increases up to about 25% from about 2%—3%. Germany also thinks highly of the direct-fired power generation, where the application of combining heat and power (CHP) with biomass is common. As the number one power country in the world, the United States also attaches a great importance to the development of bioenergy. In as early as 1991, The United States Department of Energy had a biological power generation plan and the first training region in its regional biomass energy program began in as early as 1979. Now, in the United States, the use of biomass power generation has become a choice of a large number of industrial production of electricity, and the such immense power production is used by the U.S., as the basic generating capacity in its the existing distribution system. At present, the United States has already more than 350 biomass power plants, but the U.S. Department of Energy has also put forward a development plan to gradually improve green power, and it is estimated that, by 2010, the United States will add about 11 million kilowatts installed capacity of biomass power generation 【7】.

At the present, the electricity generation with biomass is just only 1% of the entire electricity generation in the western industrial countries, but by the end of 2020, it will be 15%. By then, the West will have 0.1 billion households using the electricity from biomass power generation and the biomass power generation industry will also provide 0.4 million job opportunities.

5. Concluding remarks

The biomass energy technology research and development has become one of the world major topical issues. In the world today, many countries have developed a corresponding development plan, which forms such a development model that each of the varieties has its particular characteristics, the industry scale is expanding continuously, the technical level is increasing gradually and a good development prospect has been shown 【8】.

4.4.2　Specialized English Words

　　distributed power supply　分散的独立电源
　　suited to local conditions　因地制宜
　　traditional fossil fuels　传统的矿物燃料
　　direct-fired biomass for electric power generation　物质直燃式发电
　　traditional renewable energy sources　传统的可再生能源

biomass-rich regions　生物质能丰富的地区
large-scale food processing enterprises　大型粮食加工企业
biomass forming particles fuel　生物质成型颗粒燃料
resources dispersed rural areas　资源分散的农村地区
decentralized fuel consuming problem　分散用能问题
bio-based product　生物基产品
sunrise industry　朝阳产业
resource efficient utilization　资源高效利用
multi-joint and coordinate within the cross-course，cross-section and cross-profession　跨学科、跨部门、跨行业的联合与协同
burning wood residue　燃烧木材剩余物
combining heat and power（CHP）with biomass　生物质热电联（产）
construction phase　建设阶段
training region　实验区
job opportunity　就业机会
make use of biomass for power generation　用生物质能发电
million tons of standard coal　百万吨标煤
installed capacity　装机容量
run stably　稳定运行
corn stalk　玉米秸秆
Industrial Revolution　工业革命
over-all planning and all-round considerations　统筹兼顾
major grain-producing areas　粮食主产区
large-scale wood processing plant　大型木材加工厂
biogas technology　沼气技术
distributed resource　分散的资源
biomass industry　生物质产业
bio-energy　生物能源
biomass materials　生物质原料
expand agricultural function　拓展农业功能
contradiction between...　之间的矛盾
lagging production pattern　落后生产模式
regional power stations　区域供电站
direct-fired biomass power generation　生物质直燃式发电
planning and designing phase　规划设计阶段
regional biomass energy program　区域性生物质能源计划

4.4.3　Notes

【1】Therefore，being suited to local conditions，to make use of biomass for power generation and to transmission to the power grid or used as a distributed power supply, are good ways of turning the waste into the useful things and of saving energy, reducing environmental pollution, benefiting to the farmers.

该句较长，因此在汉译时常将其译为两句。句中由现在分词 being 引导的现在分词短语 being suited to local conditions 为全句的条件状语。句子的主语是两个并列的动词不定式 to make use of biomass for power generation 和 to transmission to the power grid or used as a distributed power supply。而表语 good ways 有两个介词短语 of turning the waste into the useful things 和 of saving energy, reducing environmental pollution, benefiting to the farmers 做其后置定语。因此整句译为："因此，因地制宜地利用生物质能发电，是一项可向电网送电，或者以分散的独立电源的方式向外供电，因而是变废为宝的好办法。这也是一项节约能源、减少环境污染、造福于农民的好办法"。

【2】This project electricity generation capacity in design is 160 million kilowatts, while the electricity generation was up to 229 million kilowatts in 2007, and counting according to the installed capacity 25 million watts, the number of operation hours throughout the year was up to 9160 hours, which reached the world advanced level and created a miracle in electricity generation with biomass.

该句也属一难长句。其主句是：This project electricity generation capacity in design is 160 million kilowatts。而由从属连词 while 引导的状语从句 while the electricity generation was up to 229 million kilowatts in 2007, and counting according to the installed capacity 25 million watts, the number of operation hours throughout the year was up to 9160 hours, which reached the world advanced level and created a miracle in electricity generation with biomass 做其状语，说明一种伴随状况。只不过该句本身又是一个很长的复合句，其中主句是：the electricity generation was up to 229 million kilowatts in 2007, and counting according to the installed capacity 25 million watts, the number of operation hours throughout the year was up to 9160 hours。其为两个并列子句 the electricity generation was up to 229 million kilowatts in 2007 和 counting according to the installed capacity 25 million watts, the number of operation hours throughout the year was up to 9160 hours 由并列连词 and 连接成为一个并列复合句，只不过在第二个子句中还有一个分词短语 counting according to the installed capacity 25 million watts 做其状语。而由从属连词 which 引导的非限定性定语从句 which reached the world advanced level and created a miracle in electricity generation with biomass 在该状语从句中做主句的定语，只不过该句中的谓语是两个并列的动词 reached 和 created。经过这样的分析，不难看出，本句应译为："该项目设计年发电能力1.6亿 kWh，2007年发电量达到了2.29亿 kWh，按2.5万 kWh 装机容量计算，全年利用小时数高达9160小时，达到了世界先进水平，创造了生物质能发电的奇迹"。

【3】The company investment project in Wangkui, Heilongjiang Province, is the first biomass power generation project in the world that uses corn stalk for the main fuel, which was formally completed and put into production on November 16, 2007, since that time, it has been running stably, and which opens up a new way of corn and the such yellow straws comprehensive utilization, for the Northeast and Central Plains region in China and has contributed to the development of direct-fired biomass for electric power generation industry for our country and the world.

该句又是一个难长句，其整个句子的框架是一个主从复合句。其主句为：The company investment project in Wangkui, Heilongjiang Province, is the first biomass power generation project in the world that uses corn stalk for the main fuel，只不过该主句并不是一个简单句，其中 that uses corn stalk for the main fuel 是修饰 biomass power generation project 的定语从句。而整个句子的从句则为由两个并列的从属连词 which 引导的非限定性定语从句，一个是：which was formally completed and put into production on November 16, 2007, since that time, it has been running stably，另一个是：which opens up a new way of corn and the such yellow straws comprehensive utilization, for the Northeast and Central Plains region in China and has contributed to the development of direct-fired biomass for electric power generation industry for our country and the world。别看第一个从句不长，但它却是一个复合句，since that time, it has been running stably 就是它的时间状语从句。而第二个从句虽然很长，但它却仍然是一个简单句（所以句子的简单与复合形式并不在于其句子的长短），只不过其谓语有两个动词并列而已，这两个并列的动词分别是 opens up 和 has contributed to，且第一个谓语动词还有一个修饰它的目的状语 for the Northeast and Central Plains region in China，所以句子显得很长。经过这样的分析，不难看出，本句应译为："该公司投资建设的黑龙江省望奎项目，是世界上第一个以玉米秸秆为主要燃料的生物质能发电项目。该项目自2007年11月16日正式建成投产以来，至今一直稳定运行，为东北地区及中原地区玉米等黄色秸秆的综合利用开辟了一条新途径，为我国乃至世界生物质能直燃式发电产业的发展作出了贡献"。

【4】In the biomass-rich regions, such as the major grain-producing areas, large-scale food processing enterprises, forest, large-scale wood processing plants and the like, a number of biomass power stations can be building according to the local conditions, while in most of the resources dispersed rural areas, with solving the fuel problem, the use of biomass forming particles fuel or biogas technology should be promoted, to solve the problem of decentralized fuel consuming with the distributed resources.

该句又是一个难长句，其整个句子的框架是一个主从复合句。其主句为：In the biomass-rich regions, such as the major grain-producing areas, large-scale food processing enterprises, forest, large-scale wood processing plants and the like, a number of biomass power stations can be building according to the local conditions，其中句干为：a number of biomass power stations can be building according to the local conditions，而 In the biomass-rich regions, such as the major grain-producing areas, large-scale food processing enterprises, forest, large-scale wood processing plants and the like 是一个做地点状语的介词短语。其从句为：while in most of the resources dispersed rural areas, with solving the fuel problem, the use of biomass forming particles fuel or biogas technology should be promoted, to solve the problem of decentralized fuel consuming with the distributed resources，其为由从属连词 while 引导的时间状语从句。应该主意的是，虽然该句在形式上是时间状语，或者说其语法功能是时间状语，但其含义却是表示一种对比或反衬，说明的是另一种情形，或者另一方面的问题。因此该句应译为："在生物质能丰富的地区，如粮食主产区、大型粮食加工企业、林区、大型木材加工厂等，可因地制宜地建设一些生物质能发电站，而在大部分资源分散的农村地区，应结合解决农村用能问题，推广应用生物质能成型颗粒燃料技术或沼气技术，实现利用分散的资源解决分散用能问题"。

【5】It is worth note that, as a new type industry, to multi-joint and coordinate within the cross-course, cross-section and cross-profession and to reconcile the contradiction between the fuel demand and lagging production patterns in rural areas are the issues requiring urgent solution.

· 151 ·

句中 It is worth note that 是强调句型，而要强调的内容则是由 that 引导出的主语从句 as a new type industry, to multi-joint and coordinate within the cross-course, cross-section and cross-profession and to reconcile the contradiction between the fuel demand and lagging production patterns in rural areas are the issues requiring urgent solution，只不过该主语太长，因此在主语句中用先行词 It 替代之，所以主句的句干便变成 It is worth note that，显得极为简洁。而从句中的主语则由两个动词不定式并列而成，其中一个是 to multi-joint and coordinate within the cross-course，cross-section and cross-profession，而另一个是 to reconcile the contradiction between the fuel demand and lagging production patterns in rural areas。因此该句译为："值得注意的是，作为一种新型产业，跨学科、跨部门、跨行业的联合与协同，以及燃料需求和农村落后生产模式之间矛盾的调和，都是亟待解决的问题"。

【6】By the end of 2004, the biomass power generation capacity in the world had reached 39 000 million watts, about 200 billion kWh in electricity generation in a year, which can substitute 70 million tons of standard coal and is the sum quantity of the generating electricity in the wind power, photovoltaic, geothermal and other renewable energy.

该句的主句是：By the end of 2004, the biomass power generation capacity in the world had reached 39 000 million watts, about 200 billion kWh in electricity generation in a year，其中 about 200 billion kWh in electricity generation in a year 是介词短语做句子的状语，对句子的谓语做进一步的说明。而 which can substitute 70 million tons of standard coal and is the sum quantity of the generating electricity in the wind power, photovoltaic, geothermal and other renewable energy 则是整个句子中的一个非限定性定语从句，其进一步说明主句中所说的那件事情。只不过该从句虽说是简单句，但其主语有两个并列的谓语动词，一个是 can substitute，另一个便是 is，所以显得较长。因此该句应译为："截至2004年年底，世界生物质能发电装机已达3900万千瓦，年发电量约2000亿千瓦时，可替代7000万吨标准煤，是风电、光电、地热等可再生能源发电量的总和"。

【7】At present, the United States already has more than 350 biomass power plants, but the U.S. Department of Energy has also put forward a development plan to gradually improve green power, and it is estimated that, by 2010, the United States will add about 11 million kilowatts installed capacity of biomass power generation.

该句的主框架是一并列复合句，其由两个并列连词 but 和 and 将 the United States already has more than 350 biomass power plants、the U.S. Department of Energy has also put forward a development plan to gradually improve green power 和 it is estimated that, by 2010, the United States will add about 11 million kilowatts installed capacity of biomass power generation 这三个并列的子句复合而成，表示这三者的关系是并列的。只不过第三个子句本身就是一个主从复合句，其主句为 it is estimated，而 the United States will add about 11 million kilowatts installed capacity of biomass power generation 是由 that 引导的主语从句。故全句应译为："目前美国已有350多座生物质能发电站，而美国能源部又提出了逐步提高绿色电力的发展计划，预计到2010年，美国将新增约1100万千瓦的生物质能发电装机容量"。

【8】In the world today, many countries have developed a corresponding development plan, which forms such a development model that each of the varieties has its particular characteristics, the industry scale is expanding continuously, the technical level is increasing gradually and a good development prospect has been shown.

该句的主框架又是一并列复合句,其由两个并列连词 and 将 many countries have developed a corresponding development plan, which forms such a development model that each of the varieties has its particular characteristics、the industry scale is expanding continuously、the technical level is increasing gradually 和 a good development prospect has been shown 这四个并列子句复合而成。只不过第一个子句本身就是一个主从复合句,其主句为 many countries have developed a corresponding development plan, 而由 which 引导的非限定性定语从句 which forms such a development model that each of the varieties has its particular characteristics 非限定性地修饰其主句, 只不过该定语从句中又含有一个由 that 引导的同位语从句 that each of the varieties has its particular characteristics, 其修饰其前面的名词 a development model, 与之为同位语关系。因此全句应译为:"在当今世界上,许多国家都制订了相应的发展计划,形成了各具特色的发展模式,产业规模持续扩大,技术水平逐步提高,呈现出良好的发展前景"。

4.4.4 Translation

生物质能发电纵论

在可再生能源发电中,生物质能发电具有电能质量好、可靠性高、技术比较成熟的特点,并可做到无间歇发电。当前,不少地区的秸秆不能得到有效利用,只能被烧掉,造成了大范围的烟气污染。因此,因地制宜地利用生物质能发电,是一项可向电网送电,或者以分散的独立电源的方式向外供电,因而是变废为宝的好办法。这也是一项节约能源、减少环境污染、造福于农民的好办法。发展生物质能发电,还可促进循环经济发展,有利于建设资源节约型、环境友好型社会,有利于社会主义新农村建设。目前,世界上大多数国家均将生物质能作为可替代能源的一个重要方面。

1. 我国发展生物质能发电的必要性与可行性

我国生物质能资源非常丰富,全国生物质能的理论资源总量接近 15 亿吨标准煤。到 2020 年,生物质能开发利用量达到 5 亿吨标准煤,这就相当于增加了 15% 以上的能源供应。与传统的矿物燃料相比,生物质能属于清洁能源,其燃烧后二氧化碳排放属于自然界的碳循环,不会对自然界形成新的污染。生物质能含硫量极低,仅为 3‰,还不到煤炭含硫量的 1/4。发展生物质能发电,实施煤炭替代,可显著减少二氧化碳和二氧化硫排放,产生巨大的环境效益。

我国农作物播种面积有 18 亿亩,年产生物质约 7 亿吨,除部分用于造纸和畜牧饲料外,剩余部分均作为燃料消耗掉了。另一方面,我国现有森林面积约 1.75 亿公顷,森林覆盖率为 18.21%,每年可获得的生物质资源量约 8~10 亿吨,所有这一切都是我国发展生物质能发电产业的良好基础。

2. 我国生物质能发电的现状

我国十分重视生物质能发电,在政策的引导下,我国生物质能发电产业得以发展迅速。最近几年来,国家电网公司、五大发电集团等大型国有、民营以及外资企业纷纷投资参与我国生物质能发电产业的建设运营。

在生物质能发电产业发展过程中,最为突出的是国家电网公司旗下的国能生物发电有限公司。该公司目前在全国 10 多个省份已核准生物质能发电项目 39 个。其中,已建成投产项目 10 个,投产装机容量 25 万千瓦;在建项目 9 个,在建装机容量 10.8 万千瓦。2007 年消耗农林废弃物约130多万吨,为当地农民增收近3亿多元,替代标煤60多万吨,减排二氧化碳60万吨以上。山东单县生物质能发电工程1×2.5万千瓦机组于2006年12月1日正式投产,开创了国内生物质能直燃式发电的先河,开辟了惠农的新途径。该项目设计年发电能力1.6亿千瓦时,2007年发电量达到了2.29亿千瓦时,按2.5万千瓦装机容量计算,全年利用小时数高达9160小时,达到了世界先进水平,创造了生物质能发电的奇迹。该公司投资建设的黑龙江省望奎项目,是世界上第一个以玉米秸秆为主要燃料的生物质能发电项目。该项目自2007年11月16日正式建成投产以来,至今一直稳定运行,为东北地区及中原地区玉米等黄色秸秆的综合利用开辟了一条新途径,为我国乃至世界生物质能直燃式发电产业的发展作出了贡献。

3. 我国生物质能发电的前景

生物质能是传统的可再生能源。在工业革命以前,一直是支撑人类社会发展的主要能源。由于矿物能源的出现,目前生物质能利用量已经很小。但随着能源资源和能源环境问题的日益突出,生物质能利用又重新成为人类的选择。若能运用先进的科学与技术,生物质能发电将部分替代目前由消耗能源矿物而生产的电能。因此,生物质能的利用要统筹兼顾。在生物质能丰富的地区,如粮食主产区、大型粮食加工企业、林区、大型木材加工厂等,可因地制宜地建设一些生物质能发电站,而在大部分资源分散的农村地区,应结合解决农村用能问题,推广应用生物质能成型颗粒燃料技术或沼气技术,实现利用分散的资源解决分散用能问题。

对于生物质能发电产业,我国政府给予了有力支持。值得注意的是,作为一种新型产业,跨学科、跨部门、跨行业的联合与协同,以及燃料需求和农村落后生产模式之间矛盾的调和,都是亟待解决的问题。如果克服这些困难,生物质能发电将在现代化建设中发挥更大的作用。

4. 国外生物质能发电产业的发展状况

自1990年以来,生物质能发电在欧美许多国家开始大发展,特别是2002年约翰内斯堡可持续发展世界峰会以来,生物质能的开发利用正在全球加快推进。截至2004年年底,世界生物质能发电装机已达3900万千瓦,年发电量约2000亿千瓦时,可替代7000万吨标准煤,是风电、光电、地热等可再生能源发电量的总和。

芬兰是欧盟国家中利用生物质发电最成功的国家之一,目前生物质发电量占本国发电量的11%。奥地利成功地推行了建立燃烧木材剩余物的区域供电站的计划,生物质能在总能耗中的比例由原来2%～3%激增到约25%。德国对生物质能直燃发电也非常重视,在生物质能热电联产应用方面很普遍。作为世界头号强国,美国也十分重视生物能源的发展,美国能源部早在1991年就提出了生物发电计划,而美国能源部的区域性生物质能源计划的第一个实验区早在1979年就已开始。如今,在美国利用生物质发电已经成为大量工业生产用电的选择,这种巨大的电力生产被美国用于现存配电系统的基本发电量。目前美国已有350多座生物质能发电站,而美国能源部又提出了逐步提高绿色电力的发展计划,预计到2010年,美国将新增约1100万千瓦的生物质能发电装机容量。

到2020年年底,西方工业国家15%的电力将来自生物质能发电,而目前生物质发电只占整个电力生产的1%。届时,西方工业国家将有1亿个家庭使用的电力来自生物质能发电,生物质能发电产业还将为社会提供40万个就业机会。

5. 结束语

生物质能技术的研究与开发已成为世界重大热门课题之一。在当今世界上,许多国家都制定了相应的发展计划,形成了各具特色的发展模式,产业规模持续扩大,技术水平逐步提高,呈现出良好的发展前景。

Unit 4.5　Hybrid PV-Battery-Diesel Power System

4.5.1　Text

　　Presently, the electrical energy demand is far greater than ever before in both developed and developing nations. Broadly speaking, the sources of conventional means of electricity generation are finite and fast depleting. Moreover, the burning of fossil fuels is the principal cause of unprecedented air pollution and environmental warming. In the wake of the above alarming issues, as an option to avert/mitigate impending energy crisis, most nations worldwide are prompted to embark on research on solar energy.

　　The resource of fossil fuel is decreasing day and day. At the same time, the awareness of the important of protecting the global environment is growing. So the searching for the pollution-free replaces of fossil resources becomes more and more urgent. Photovoltaic source as renewable energy is pollution-free and inexhaustible. So the photovoltaic generation technology gets more and more attention. With pries reductions of PV modules and development of PV technology, the role of PV generation systems is gradually changed from the supplemental energy to the substitute energy. This thesis presents the configuration, the operation principle and the design procedures of a low power photovoltaic inverter connected with electric utility lines, and deals with such key problems as true maximum power point tracking(MPPT)of PV array, improvement of inverter out put wave, islanding effete and so on by establishing related simulation models and applying novel control strategies and experiment validations 【1】.The main contents of the paper are as follows:

　　(1) The actuality and foreground of PV sources abroad or our country is introduced. The basic structure and types of the PV source also are presented, holding apprehend in the mass to the system.

　　(2) This paper has studied on the basic principle and output characters of PV cell. The emphasis is on the output characters and its effect factor.

　　(3) The output of PV is non-linear, which has Maximum Power Point tracking (MPPT) problem. This paper has studied MPPT method and proposed improved measures.The method of tracing the maximum PV powered points is realized with a boost converter in this design. The circuit is experimented by the emulator PSIM at last. And then the thesis explains the operation principle and control strategy of boost converter, the components parameters of the main circuit are introduced and the design procedure of the control circuit is given. The boost converter is controlled by a core P89LPC938 and TL94.

　　(4) The operational principles of photovoltaic grid-connected generation system are analyzed. The topology structure of the inverter section is single-phase full bridge, whose operation principle, control strategy, hardware and software design are presented in this thesis. The control method is hysteretic tracking method and the DC-AC inverter is controlled by DSP TMS320LF2407. The islanding phenomenon of grid-connected PV system is discussed, which harm and detection

methods are investigated. Simulation results verify that method using PSIM.

Solar energy is one of the inexhaustible, clean (does not produce emissions that contribute to the greenhouse effect), and potential source of renewable energy options. Solar collectors can be classified as either solar thermal energy converters or solar electric energy converters. The devices that directly convert solar into electric energy are generally called photovoltaic (PV). The concept of PV is well understood and currently (in spite of hindrances) thousands of PV-based power systems are being deployed worldwide, for providing power to small, remote, grid-independent or stand-alone applications. The use of renewable sources of energy reduces combustion of fossil fuels and the consequent CO_2 emissions, which is the principal cause of global warming. Global warming is expected to change terrain and climate of many countries unless measures are taken. More importantly, in the light of December 1997s Kyoto protocol on climate change (due to carbon emissions), about 160 nations have reached a first ever agreement (to turn to renewable/wind/PV power) to limit carbon emissions, which are believed to cause global warming. Although, the solar energy is enormous, but PV-driven power system is still an expensive option(PV system capital cost is about 4000 \$/kW, while the capital cost of conventional power systems is about 1000 \$/kW). The high initial investment cost in PV systems is still the main road-block that hinders promotion of this technology in large-scale applications. Nonetheless, PV finds application in remote areas (where it is uneconomical to extend the utility grid) not served by an electric grid. PV systems have the advantage of minimum maintenance and easy expansion (upsizing) to meet growing energy needs. PV modularity (modules are available off-the-shelf) is one of its major strength and it allows the users to tailor PV system capacity to the desired situation 【2】. PV systems also produce electricity during the times when we demand it most, on hot sunny days coinciding with our peak electricity consuming periods. The demerits are: PV is capital-cost-intensive and its sunshine-dependent output does not match the load on 24-h basis. However, major technological milestones (yielding cost reduction of PV, improved efficiency, etc.) may change the scenario.

The solar insulation varies not only during different seasons but also at different times of the day. Therefore, for applications where energy is required for a 24-h period, the need cannot be met with a PV system alone. In this condition, the integration of PV installations with the battery storage or diesel system or with both, can meet the required load distribution on a 24-h basis.

Despite abundant availability of solar energy, a PV system alone cannot satisfy load on a 24-h basis. Stand-alone diesel (relatively inexpensive to purchase) are generally expensive to operate and maintain especially at low load levels. Often, the variations of solar energy generation do not match the time distribution of the demand. Therefore the power generation systems dictate the association of battery storage facility to dampen the time-distribution mismatch between the load and solar energy generation. PV-generated electricity stored in batteries can be retrieved during nights. The use of diesel system with PV-battery reduces battery storage requirement. The research carried out in other parts of the world indicate that hybrid combination of PV/diesel/battery system is a reliable source (reliability is one of the main selling point) of electricity.

The hybrid PV-battery-diesel configuration, by virtue of a high degree of flexibility, offers several advantages. The diesel efficiency can be maximized, the diesel maintenance can be minimized and a reduction in the diesel and battery capacities while matching the peak loads can occur 【3】. The present investigation shows that the potential of renewable energy option of solar energy cannot be overlooked 【4】. The observations of this study can be employed as a benchmark in designing/sizing of hybrid PV diesel battery systems for other locations having similar climatic and load conditions.

Over dependence on fossil fuels is alarming. Hence, investments in solar energy are imperative to mitigate energy crisis in foreseeable future.

4.5.2 Specialized English Words

hybrid PV-battery-diesel power system　太阳能光伏电池与柴油机发电机的混合发电系统
PV（photovoltaic）　光伏
photovoltaic source（or solar energy）太阳能
PV module　光伏模块
PV generation systems　太阳能光伏发电系统
photovoltaic inverter　光伏逆变器
operation principle　运行原理
PV array　光伏阵列
inverter out put wave　逆变器的输出波形
islanding effete　孤岛效应
boost converter　升压变换器
single-phase full bridge　单相全桥电路
solar collectors　太阳能集热器
solar electric energy converter　太阳能电力转换器
coinciding with　恰逢
time distribution of demand　需求在时间上的分布
conventional means of electricity generation　传统发电方式
air pollution and environmental warming　空气污染和气候变暖
impending energy crisis　迫在眉睫的能源危机
protecting global environment　全球的环境保护
supplemental energy　补充能源
substitute energy　替代能源
electric utility line　公共输电线（电网）
design procedure　设计方法
true maximum power point tracking（MPPT）　真正的最大功率点跟踪
novel control strategy　新型控制策略
effect factor　发电效率
topology structure　拓扑结构
hysteretic tracking method　滞环跟踪法
solar thermal energy converter　太阳热能转换器
Kyoto protocol　京都议定书

variations of energy generation　发电量的变化

peak loads　峰值负荷

4.5.3　Notes

【1】This thesis presents the configuration, the operation principle and the design procedures of a low power photovoltaic inverter connected with electric utility lines, and deals with such key problems as true maximum power point tracking (MPPT) of PV array, improvement of inverter out put wave, islanding effete and so on by establishing related simulation models and applying novel control strategies and experiment validations.

应该看出尽管该句较长，但其仍然只是一个简单句，主语为 This thesis，谓语由两个并列的动词 presents 和 deals with 组成。只是第一个谓语动词的宾语较长，其由三个并列的名词、名词词组和名词性短语组成。具体地说，第一个是名词 the configuration，第二个是名词词组 the operation principle，而第三个则是名词性短语 the design procedures of a low power photovoltaic inverter connected with electric utility lines。第二个谓语动词的宾语虽然较短，只有 such key problems 寥寥数语，但其后又有一个由介词 as 引导的介词短语 true maximum power point tracking (MPPT) of PV array, improvement of inverter out put wave, islanding effete and so on 做其定语。不仅如此，其后又有一个介词短语 by establishing related simulation models and applying novel control strategies and experiment validations 做第二个谓语动词的状语，因此整个句子显得较长。但经此分析，不难看出，本句应译为："本文提出了一种小功率的光伏并网逆变电源的具体结构、运行原理和设计方法，并通过建立有关的仿真模型，以及采用新型控制策略和实验验证等手段，解决了诸如光伏阵列的真正的最大功率点跟踪、逆变器的输出波形的改善，以及并网中的孤岛效应等这样一些关键问题"。

【2】PV modularity (modules are available off-the-shelf) is one of its major strengths and it allows the users to tailor PV system capacity to the desired situation.

该句是一并列复合句，虽然其并列连词 and 在语法上的作用是并列两个子句，但其含意在此处则是表结果的。因此，该句应译为："光伏发电系统的模块化（也就是集成在一块板上的模块）是它的一个主要的优点，这样就可使用户根据容量需要拼接出自己想要的发电系统"。

【3】The diesel efficiency can be maximized, the diesel maintenance can be minimized and a reduction in the diesel and battery capacities while matching the peak loads can occur.

从总体上说，该句是一个并列复合句。只是需要看清楚的是第三个子句的主语是 a reduction，而谓语是 can occur，且其间离得较远。此外，整个介词短语 in the diesel and battery capacities while matching the peak loads 是 a reduction 的后置定语。这样看来该句就应翻译为："柴油机的效率可最大化，柴油机的维修工作可最小化，与峰值负荷相匹配的柴油机和蓄电池的容量可减小"。

【4】The present investigation shows that the potential of renewable energy option of solar energy cannot be overlooked.

该句为一形式否定句。所谓形式否定就是其在形式上是否定的，但其意则是肯定的，且其肯定之口气更加强硬。这种句型多出现于有 cannot 等有"否定"意义的词和 over 等有"过分"意义的词同时出现的句子中。究其原因便是 cannot 是"做不到"的意思（注意：一看到

cannot，中国学生马上就译为"不能"。然而"不能"在中文中是"不可以"的意思，这样就将其意正好翻译反了。像本句就会翻译成："不能看得过分的高"，这一点应该引起中国学生的足够重视。应该说，这种语法现象中国学生不是没见过，也不是没有训练过，但为什么常常在此犯错误，究其原因便是对这种语法现象知其然，而不知其所以然。老师们也往往说这是一种习惯，就这么翻译，好像没有什么道理，其实不然。究其原因便是中国学生没有真正弄懂 cannot 的意思，"做不到，不可能"才是其真正意思、根本的意思、原始的意思）。因此本句应翻译为："本文的研究表明，太阳能这种可再生能源的开发利用潜力，再怎么看高都不过分"。

4.5.4 Translation

太阳能光伏电池与柴油机发电机的混合发电系统

　　目前，无论是在发达国家还是发展中国家，对电力能源的需求都远远大于以往任何时候。总体说来，传统发电方式所使用的能源是有限的，并且正在日益枯竭。更何况，前所未有的空气污染和气候变暖等环境问题的主要根源是燃烧矿物燃料。在对上述令人担忧的问题警醒后，作为一种避免或减少迫在眉睫的能源危机的一种选择，全世界大多数国家都奋起着手进行太阳能的研究。

　　矿物燃料资源正在日渐减少，而与此同时，人们对全球环境保护问题的重要性的认识程度也在不断提高。因而寻找无污染的可替代矿物燃料的能源问题变得越来越迫切。太阳能作为一种可再生能源，洁净无污染，并可持续利用，因而太阳能发电技术越来越得到人们的重视。随着光伏模块价格的不断降低和光伏技术的不断发展，太阳能光伏发电系统的作用，也由现在的处于补充能源地位，逐渐地转向替代能源的地位。本文提出了一种小功率的光伏并网逆变电源的具体结构、运行原理和设计方法，并通过建立有关的仿真模型，以及采用新型控制策略和实验验证等手段，解决了诸如光伏阵列的真正的最大功率点跟踪、逆变器的输出波形的改善，以及并网中的孤岛效应等这样一些关键问题。本文的主要内容如下：

　　（1）介绍了目前国内外光伏发电的现状和发展前景，并介绍了光伏发电系统的结构组成和类型，对光伏发电系统给出了一个总体认识。

　　（2）研究了光伏电池的基本发电原理和输出特性，重点是其输出特性及发电效率。

　　（3）光伏电池输出特性是非线性的，这就有一个输出最大功率跟踪的问题。本文研究了跟踪最大功率点的方法，并提出了改进方案。该设计方案采用了升压变换器，从而达到了跟踪光伏电池的最大功率点的目的。最后运用 PSIM 仿真软件进行仿真验证分析。随后，本文又分析了升压变换器的工作原理和控制策略，给出了电路的参数选择，介绍了控制电路的设计过程。该部分控制核心芯片是 P89LPC938 和 TL94。

　　（4）对光伏并网发电系统的运行原理进行研究。系统中变换器部分的拓扑结构是单相全桥电路，其工作原理、控制策略、软硬件的设计文中也已介绍，其控制方法采用输出电流滞后跟踪方式，其直流至交流的变换器的控制核心芯片为 TMS320LF2407。对并网供电系统在工作过程中出现的"孤岛效应"现象、其危害性和产生的原因进行了探讨。PSIM 软件仿真的结果证明了该方法的有效性。

　　太阳能是一种取之不尽、用之不竭的，清洁的（不产生有助于温室效应的气体排放），以及潜在的可再生能源。太阳能集热器可被定义为太阳热能转换器或太阳能电力转换器。一般将太阳能直接转换成电能的系统称为光伏（太阳能）系统。光伏发电这个概念很好理解，虽然还存在着一些障碍，但目前在全世界已有成千上万的太阳能光伏电源系统正在建设之中，其为小型的、偏远的、未连网的，或是孤立的用户提供电力。利用可再生能源可以减少燃烧矿物燃料和由此产生的二氧化碳排放量，这是全球气候变暖主要的原因。除非采取措施，否则全球变暖将改变许多国家的地势和气候。更为重要的是，1997年12月在日本签订了关于二氧化碳的排放会引起气候变化的《京都议定书》，在其指导下，大约160个国家就转向可再生能源的利用，大力发展风力和光伏发电，以限制二氧化碳排放量（二氧化碳的排放被认为会引起全球变暖），已经达成了有史以来的第一次协议。尽管太阳能的能量是巨大的，但光伏发电系统仍然是比较昂贵的（光伏发电系统的资金成本约4000美元／千瓦，而常规的电力系统的资金成本才约1000美元／千瓦）。初期的高投资成本，仍然是阻碍开展这一技术大规模地进行应用的主要绊脚石。但无论怎么说，光伏发电在边远地区还是找到了其应用，这些地区因经济不发达而无力发展供电电网系统，故不能取用电网的电能。光伏发电系统的优点是维护简单和易于扩建（改建），以满足日益增长的能源需求。光伏发电系统模块化（也就是集成在一块板上的模块）是它的一个主要优点，这样就可使用户根据容量需要拼接出自己想要的发电系统。它使用户在理想的状况下能够定量定制光伏系统的能力。光伏发电系统在我们对电能需求最大的时候，也就是在最炎热的大白天恰逢用电高峰时也能发电。光伏发电系统的缺点是投资成本太高，且高度依赖于日照，因而不能满足24小时不间断的用电需要。然而，重大的技术里程碑性的突破（包括光伏太阳能成本的降低，以及效率的提高等）可能会使情况发生变化。

太阳的光照不仅在不同的季节里是变化的，而且在一天中的不同时段也是变化的。因此，在那些需要24小时连续供电的地方，光靠光伏发电系统是不能满足要求的。在这种情况下，将光伏与蓄电池或柴油发电机组，或是其二者组成一体化的设施，便可达到向负载24小时连续供电的需要。

尽管太阳能极为丰富，但光伏发电系统系统本身并不能满足负载24小时的供电需求。独立的柴油机在购买时相对便宜一些，但运行和维护费用一般昂贵，特别是在低负荷的情况下更是如此。通常的情况是，太阳能光伏发电量的变化，与负载的需求在时间上的分布是不相匹配的。因此，光伏发电系统便给予与之相互配置的蓄电池设备下指令，令其弥补光伏太阳能发电系统的发电时段与负载用电分布时段之间不匹配的这段间隙。储存在蓄电池中的太阳能光伏发电所发的电能在晚间可取出。运用柴油发电机系统可降低对蓄电池的容量的要求。还有一些地区的研究也表明，光伏/柴油/蓄电池混合发电系统是一种可靠的电源，其可靠性是其一个主要的卖点。

光伏蓄电池柴油机混合发电系统凭借其高度的灵活性，拥有若干多个优点。柴油机的效率可最大化，柴油机的维修工作可最小化，与峰值负荷相匹配的柴油机和蓄电池的容量可减小。本文的研究表明，太阳能这种可再生能源的开发利用潜力，再怎么看高都不过分。本文的研究结果，可作为其他在具有类似的气候条件，以及类似的负载情况之处设计，或者规划光伏蓄电池柴油机混合发电系统的样板。

过度依靠矿物燃料的警钟正在时时敲响。因此大力开展太阳能利用的研究，以应对即将出现的能源危机已迫在眉睫。

Part 5 Computer & Artificial Intelligence

Unit 5.1   Computer and Its Hardware

Unit 5.2   Computer Software and Human Resources

Unit 5.3   Personal Computer Word Processing and Electronic Spreadsheets

Unit 5.4   Intelligent Technology

Unit 5.5   Neural Nets

Unit 5.1 Computer and Its Hardware

5.1.1 Text

A computer is a device that, under program control, performs arithmetic and logical operations without human intervention. The computer has proliferated because it can process data and deliver information efficiently, effectively, and at relatively low costs. In general, a device can be classified as a computer if it can:

(1) Perform arithmetic operations on data, including the basic functions of addition, subtraction, multiplication, division, and (under control of special devices or features) some higher-level operation.

(2) Perform logical operations on data. A computer can compare two data items to determine whether one is equal to, large than, or smaller than the other 【1】.

(3) Be programmed to operate without human intervention. Sequences of instructions to perform input, arithmetic, logical, output, and storage operations can be written and placed inside a computer and executed automatically.

Computer Characteristics

A computer is an electronic device. It uses circuits capable of amplifying or regulating electrical current. The electronic components used in computers perform at high speed and with great reliability.

Computers are also bi-state devices. Bi-state means that only one of two electrical conditions can exist at any given time. For example, within a computer, electrical current either flows along a wire or it doesn't; electricity flows either in one direction or the other; a circuit is either open or closed to the flow of electricity; an electronic switch is set either on or off. This bi-state characteristic is fundamental to all computers.

Most computers are digital devices. Digital simply means that data are coded as numbers, or digit. Because a computer is a bi-state device. It can represent only two different digits, or data values0 and 1. For example, when electricity is allowed to flow through a circuit, this condition can be represented as the digit 1. When a circuit is closed to the flow of electricity, the condition will be represented by the digit 0. Thus, all data processed by a computer and all of the instructions that control processing must be coded and represented internally as combinations of 0 and 1 data values.

In the very earliest days of the computer era, programs had to be written in this binary language. Data also had to be translated into their binary equivalents before being entered into the computer. Today, computer still are controlled through binary-coded instructions. However, modern computers also can recognize data and instructions that resemble human-readable language. People no longer have to communicate with computers solely through use of binary language.

Computer Hardware

The four basic information processing functions—input, processing, output, and storage—correspond with the four basic units of computer hardware:

(1) Input units, or devices, capture data and make them available for processing.

(2) The processor unit contains the electronic circuitry that arithmetic and logical operations required for sorting, classifying, calculating, summarizing, and comparing data.

(3) Output units communicate or report the results of processing to people who need the information.

(4) Storage units retain data and information for future retrieval, change, and processing.

Input Devices

To provide the programs and data necessary for computer processing, two corresponding requirements also must be met. First, a method must exist for entering original programs into the computer. Second, data must be made available to the programs for processing. These functions handled by computer input units.

Programs and data originate in human-readable form. The great majority of programs are written in English—like languages understandable by people. Data also originate in human-readable form, often as handwritten or typewritten business transaction documents. Before these programs and data can be processed by a computer, they must be encoded in machine-readable form and input.

The most popular type of input device is the keyboard. A keyboard may be part of a hard-copy work station that provides printed copies of the keyed data. The actual input function occurs when the keyed characters are entered into the computer for processing.

Processor

The processor units is the heart of the computer system. Here, the actual arithmetic and logical operations are carried out. The processor unit is composed of three main functional parts:

(1) Memory
(2) Control unit
(3) Arithmetic/logic unit

Memory

Memory, also called main memory or primary storage, is a storage area that holds programs and data temporarily during processing. Whenever a program is entered into the computer from some input device, it is stored in memory until processing is completed. Under control of the program, data are likewise brought into memory and made available for processing. In a sense, memory serves as the computer's "scratch pad". It is important to keep in mind that memory is only a temporary storage area. No actual processing takes place in memory. Once the computer has completed its processing, memory is cleared that made available for use in running other programs with other data.

Control Units

The control unit contains the circuitry for directing the operations of the computer and all devices attached to it. When programs or portions of programs are input by a computer operator, the control unit directs the instructions to areas in memory to await processing. Then, the control recalls program instructions one at a time, as needed, evaluates each instruction to determine what operation is required, and activates the electronic circuitry and data paths necessary to carry out the operation.

Arithmetic/Logic Unit

The arithmetic/logic unit (ALU) performs the actual processing operations involving addition, substraction, multiplication, division, and comparison of two data values. After the control unit has set up the necessary paths, data are brought from memory to the arithmetic/logic unit and placed in registers. These registers are similar in function to the display registers found in electronic calculators. Within these registers, arithmetic and comparison operations are performed. Processing results (for example, answers developed from arithmetic calculations) are returned to memory, at which point they are made available for output or for further processing. The arithmetic/logic unit and the control unit collectively are referenced to as the central processing unit (CPU).

Computer memory must be large enough to hold both the program and are being processed at the moment. Therefore, the size of memory is an important consideration in acquiring a computer. Computer memory is measured in terms of the number of characters, or bytes, of memory that are available. This value is often expressed in K, or M, of bytes. The letter K is a standard symbol for the value thousand, while M for the value million.

On small systems with limited memory capacity, room usually is available for only one program and the data to be processed by the program. On these system, only one person at a time can use the computer. On larger systems with extensive memory capacity and very fast operating speeds, two or more people can use the computer simultaneously. These multiprograming systems divide memory into several portions. Each user is assigned a portion of memory large enough to hold his or her program and the necessary data. The computer can process all of the programs by interspersing fractions of seconds of time to each user. Because the computer operates and alternates among the different programs so quickly, each person has the impression that he or she is the sole user.

Output Devices

After processing is completed by the CPU, the results are returned to memory to await further processing, transfer to storage, or output. The main function of output devices is to transform information from machine-readable to human-readable form. Thus, output devices perform a communication function so that people can access and understand the results of processing.

Computer output can be transformed either as text (words, number, and other special symbols) or as graphics (charts, graphs, drawings, or pictures). One way to output the results of processing is on video display terminal (VDT).

For user who need a permanent, human-readable record of processing results, printing terminals, can be used.These hard-copy devices are controlled by the computer, which transmits coded information to the printer.The circuitry within the printer translates the signals into printed characters, producting a typewritten-style copy of the output.

As with input units, several different methods can be used to produce computer outputs and to convert them into human-readable form.

Storage Devices

Recall that, once a program has completed its processing, computer memory is released to support processing of other programs.The copy of the program that was in memory is written over and replaced by the next program.So, every time a program is to be run, it must be loaded into memory.Most of these programs and data files are read into memory from storage devices used for long-term retention.Programs and data can be written from memory to these devices and later reloaded into memory as needed.Original program or data entry, therefore, need take place only one time.

5.1.2　Specialized English Words

  bi-state devices　双电位状态装置
  instructions　指令
  be coded　数码化
  binary equivalent　等值的二进制数
  data sorting, classifying, calculating, summarizing, and comparing　数据的排序、分类、计算、累加及比较
  hard-copy work station　硬拷贝工作站
  register　寄存器
  hard-copy　硬拷贝
  main and functional part　主要和基本部件
  diskettes, or floppy disks　软磁盘
  read/write heads　读/写头
  main memory or primary storage　主存储器
  central processing unit（CPU）　中央处理器
  coded as numbers　数码表示
  control processing　控制程序
  binary language　二进制语言
  binary-coded instruction　二进制的代码指令
  sequences of instructions　指令系列
  without human intervention　无人干预
  ideo display terminal（VDT）　显示终端
  long-term retention　长期保存

audio or video recording tape　录音带或录像带
hard disk，or fixed disks　硬磁盘
retrieve　检索
archival　存档的，备用的
data path　数据通道
charts，graphs，drawings，or pictures　标绘图、图表、描绘图或图片

5.1.3　Notes

【1】A computer can compare two data items to determine whether one is equal to，large than，or smaller than the other.

该句为一主从复合句，由 whether 引出的宾语从句 whether one is equal to，large than，or smaller than the other 做 to determine 的宾语，其中的引导词 whether 本身又有"是……，或是……"的意思。因此全句译为"计算机能将两个数据进行比较，从而能确定其是相等、不等、大于还是小于的逻辑关系"。

5.1.4　Translation

计算机及其硬件

计算机是一种在程序控制下自动完成算术及逻辑操作的装置。因为计算机能有效地、有力地处理数据及传送信息，并且成本相对低廉，因此已得到广泛地应用。一般来说，一台具备以下功能的装置就能视其为计算机：

（1）可完成加法、减法、乘法、除法等基本算术运算（在某些专用设备或器件控制下）和某些更高级的运算。

（2）可对数据进行逻辑运算。计算机能将两个数据进行比较，从而能确定其是相等、不等、大于还是小于的逻辑关系。

（3）编程后可在无人干预下运行。指令执行程序输入、算术、逻辑、输出和存储等操作功能的指令编程能写入和置于计算机内，并能自动执行。

计算机的特性

计算机是一种电子装置，其运用能放大或调整电流的电路。计算机中运用的电子元件速度和可靠性极高。

计算机也是一种双电位状态的装置。所谓双电位状态，是指对于任何特定的时间，在两种电位状态中只能存在其中的一种。例如，在计算机内，导线中要么有电流，要么没有电流；电流流向要么是某个方向，要么是另一个方向；电路对电流在其中的流动来说要么是开路，要么是闭路；电子开关的状态要么是闭合，要么是断开。双电位状态特征是一切计算机运行的基础。

绝大多数计算机是数字式装置。所谓数字式，是指数据是用数码来表示的。因为计算机是一种双电位状态装置，所以其只能表示出 0 和 1 这两个不同的数字。例如，当电路中有电

流时，这种状态可用数字 1 表示；当电路中无电流流动时，这种状态可用数字 0 表示。因此，由计算机所处理的所有数据，以及控制程序的所有指令，在计算机内部必须数码化，并一律表示为由 0 和 1 这两个数字组成的数字组合。

在计算机运用的早期，程序都得用这种二进制语言来表示。数据在输入计算机之前也得用转换成等值的二进制数。如今，计算机仍然由二进制的代码指令来控制。然而，现代计算机也能识别类似于人类易于阅读的语言编写的数据和指令。人们不再只能用二进制语言与计算机交流了。

计算机硬件

计算机的四种基本信息处理功能——输入、处理、输出和存储是与下述计算机硬件的四种基本单元相对应的：

（1）输入单元（或装置），其功能是获取数据并使之便于处理。

（2）处理器单元，这是这样一种电子电路，其能进行诸如数据的排序、分类、计算、累加及比较等算术及逻辑运算。

（3）输出单元，为需要信息的人交流和报告处理结果。

（4）存储单元，可存储数据及信息，以便以后进行检索、修改和处理。

输入单元

要为计算机的处理过程提供必要的程序和数据，就必须同时满足两个相应的要求。第一，必须有一种将原始程序输入给计算机的方法。第二，为了便于处理，数据必须是经处理后可用于程序的。这两种功能都是由计算机输入单元来完成的。

原始的程序和数据都是源自人类易于阅读的形式。大部分程序都是用类似于英语这样一种易于人类理解的语言编写的。数据也源自人类易于阅读的形式，常常为手写或商务文件的那种打字形式。在这些程序和数据能够由计算机处理之前，其必须转换为机器可识别的形式并将其输入。

最常用的输入设备是键盘。键盘还可以说是计算机的硬拷贝部件的一部分，可提供关键数据的打印件。实际的输入功能出现在当关键的数字输入计算机中以备处理时。

处理器

处理器单元是计算机系统的心脏。具体的算术及逻辑运算就在这里进行。处理器由下列三个主要的基本部件构成：

（1）存储器

（2）控制单元

（3）算术/逻辑运算单元

存储器

存储器也称为主存储器，其为计算机处理信息期间暂时存放程序及数据的地方。一旦程序从某个输入部件输入计算机内，在处理过程完成之前，它便一直存储在主存储器中。在程序控制下，数据可能存入存储器中并使其为处理过程所用。从某种意义上说，主存储器是计算机的暂存平台。存储器仅仅是一个暂时存放之处，记住这点是很重要的。在存储器中是没有实际上的处理过程发生的。一旦计算机完成处理过程后，主存储器内的内容就会被消除，以供带有不同数据的其他程序运行使用。

控制单元

控制单元包括控制计算机及与之相连接的所有设备运行的线。当计算机操作员将程序或其中一部分输入计算机后,控制单元便将这些指令引导到存储器的某个地方等待处理。然后,控制器依次调出程序的指令,必要时还会对每条指令进行判断,以便确定进行哪一步操作,然后驱动执行这些操作所需的电子线路和数据通道。

算术/逻辑单元

算术/逻辑单元(ALU)的功能是完成加法、减法、乘法、除法和两数值的比较等实际操作。当控制器建立了所必需的通路后,数据就从存储器传送到算术/逻辑单元并存入寄存器中。这些寄存器的功能类似于电子计算器中的显示寄存器的功能。在这些寄存器中,实现算术及比较运算。其处理的结果(比如,算术运算产生的结果)存入存储器中,以便可以输出或做进一步的处理。算术/逻辑单元和控制单元合在一起称为中央处理器(CPU)。

计算机的存储器容量必须不仅要足以存储程序,还要存储数据处理所需的所有数据记录。因此,在购买计算机时,存储器容量的大小是该考虑的一个重要因素。计算机存储容量的大小一般以可用的存储器字节数表示。该值常用 K 和 M 表示,字母 K 是一千标准字符,M 是一百万标准字符。

在存储器容量有限的小型计算机中,其存储空间通常只能存储一个程序和与之有关待处理的数据。这些小型计算机,同一时间只可供一人使用。而具有大存储容量及高运算速度的大型计算机,则可供两个以上的操作者同时使用。这些多程序计算机将存储器分成几部分。根据其程序和必要的数据的大小,将某一个相应的存储模块指定给每个具体的用户。计算机通过将每个用户的时间分割成若干段,便能在几乎同一时间里处理各个用户的程序。因为计算机运行的速度,以及在每个不同程序之间交替变化的速度是如此之快,以至于每个用户都感觉自己是唯一的使用者。

输出装置

CPU 在处理完程序后,其结果被送至存储器中,或是等待下一步的处理,或是转送至存储器,或是输出。输出装置的主要功能是将机器易识别的信息转换为人易识别的信息。因此,正是由于输出装置具有这样的通信功能,人们才得以获取和了解处理的结果。

计算机的输出可转换成文本(字、数据、或其他特殊的字符)形式,也可转换成图形(标绘图、图表、描绘图或图片等)形式。输出处理结果的一种方法是将其显示在显示器(VDT)上。

对于那些需要长久保存和人类易于阅读处理结果的记录的用户来说,其可使用打印输出。这些硬拷贝的设备由计算机控制,后者将信息代码传送给打印机。打印机内部的电路将这些信息转换为打印的字符,从而生成打印形式的输出拷贝。

由于可用输入单元输入各种不同的命令,因此可有多种方法生成计算机输出并将其转换成人类易于阅读的形式。

存储设备

如前所述,一旦计算机的程序处理完成了其处理过程,其存储器就会让出其空间以供别的程序使用。存储器中原来的程序的版本会重写,并被下一个程序所取代。因此,每当要运行某一程序时,就必须将其下载到存储器中。这些程序和数据中的大多数会从可长期保存的存储设备中读入存储器。程序和数据也从存储器写入这些存储设备,随后又在需要时再返回存储器。因此,原始程序和数据的输入只需操作一次。

Unit 5.2　Computer Software and Human Resources

5.2.1　Text

Through use of software, or programs, computer hardware is activated to assist people in their work, entertainment, and day-to-day lives.

Programming Languages

A computer program is a detailed, step-by-step set of instructions that the computer must perform on data. These commands are written in special programming languages that are understandable by the computer. More than 200 different languages have been created. However, only about a dozen of these languages are in popular use.

Most computer programs are written in languages that resemble standard English and/or mathematical terms. Thus, programs can be written in human-readable form. This feature makes it convenient for people to express problem solutions in their own languages rather than in the binary language of the computer. The computer then uses a language translator program to convert these English-like instructions into machine-language for execution by the computer.

Programming Languages Types

(1) Business programming languages are designed to solve business programs. Business programs frequently require storing and processing large volumes of data. Business processing ususlly is accomplished through the application of basic arithmetic on data. Business programming languages offer features that are tailor-made to meet these processing needs conveniently and efficiently. Languages such as COBOL (COmmon Business-Oriented Language) and RPG (Report Program Generator) are popular for business information processing.

(2) Scientific programming languages are used for applications characterized by relatively low volumes of data and by sophisticated mathematical processing. Scientific, engineering, and mathematical applications require specialized calculations and language features that provide these calculations automatically. Programming languages such as FORTRAN (FORmula TRANslator) are popular with the scientific community.

(3) General-purpose programming languages are designed to combine the data handling and formatting capabilities of business languages and the mathematical processing features of scientific languages into a single programming language. These languages, such as BASIC (Beginner's All-purpose Symbolic Instruction Code), PL/1 (Programming Language One), and Pascal, are useful for almost any type of processing. General-purpose languages often are used to teach students how to program.

In addition to these types of languages, programmers make use of other languages to processing text, to produce graphics, and to perform other specialized applications.

Types of Computer Software

Computer software can be classified according to two broad categories:
(1) Application software
(2) System software.

Application software

Programs written in languages such as COBOL, RPG, BASIC, and FORTRAN are created specifically to meet business and/or scientific processing needs. The languages allow people to express their processing requirements in English-like form.

The computer cannot execute application program instructions directly. The computer only understands the binary language of zeroes and ones.So, in fact, the user's program only specifies what operations are to be performed, not how to perform them 【1】. The computer, therefore, requires instructions that tell the hardware exactly how to carry out the requested operations.

System Software

Programs that act upon instructions provided in application programs are known collectively as system software 【2】.These routines interpret user program commands contained in application programs and activate the proper hardware.Also referenced to as the computer's operating system (OS), system software comes packaged on disk, called a disk operating system (DOS), for loading into memory. On some systems, specially home and personal computers, the operating system, or portions thereof, is recorded permanently, into the internal memory circuits of the computer.

System software, then, is an intermediary between the user's applications program and the computer hardware.When the user's program specialties an input operation, for example, the system software calls upon one of its routines to activate the input device, to accept and transmit the data into memory, and to allocate storage space within memory to hold the data.

As you may have noticed, several levels of computer commands can be used. As a user, you communicate with the computer by submitting data to be processed by an application program. The data trigger processing functions built into the program. The application software, in turn, issues commands to the system software requesting specific operations. The system software, then, communicates with the computer hardware to carry out the actual operations. As a user, you are insulated from complex, highly technical consideration because of this buffering effect of software.

Human resources

Regardless of how large or sophisticated computers become, people remain the key ingredient for viable, result. producing computer systems. The scope of computer applications is so great that the people involved play specialized roles. These roles include:

(1) Users, or customers, establish needs and ultimately use the results delivered by computers.
(2) System analysts and system designers analyze needs and design computer-based systems to satisfy those needs.
(3) Programmers design, write, and maintain the applications software that implements user needs.

(4) Computer operators operate the equipment in a production environment to process data and deliver information to users.

(5) Technical specialists maintain the computer hardware, systems software, and data resources.

These and other specialists populate organizations that rely on computers, and assist in bringing computers into the service of human needs.

Users/Customers

Computer systems need customers. These customers are the people who have needs for and place demands upon computers for specific services. A computer cannot perform usually unless there is a user group that has a need for the results of processing 【3】. Although they may not have a "hands-on" need for computers, users play a critical role in defining the tasks computers perform and the products they deliver 【4】.

Because of this key role, the relationships among users and information processing professionals are important. Users must be able to communicate with computer specialists to indicate what they need, why they need it, and when end products must be made available. On the other hand, computer professionals must understand that if they did not produce useful products, they wouldn't be needed 【5】.

Mutual understanding and cooperation must be the watchwords if computers are to be applied successfully and profitably.

Systems Analysts and Designers

Within any large organization, specialists can be found who work with users to develop computer-based systems that will do the jobs and produce the results required. The most common job title in the systems development area is systems analyst. These specialists analyze business needs and develop systems to satisfy those needs. In doing so, the analyst and designer form a bridge between the day-to-day business world of computer users and the highly technical world of computer specialists.

Systems analysts work individually or in project teams with users. Together, users and analysts study the business environment and develop an understanding of the business procedures to be computerized, the data to be processed, and information to be delivered 【6】. Users and analysts develop sets of documents indicating what functions a computer system should perform. Then, these functional specifications become the blueprints for designing a computer-based system 【7】.

Systems designers work from these documents to design the actual configurations of hardware, software, data files, manual procedures, communication networks, and human resource that will be required to implement the system 【8】. Designers take the functional requirements identified during the analysis step and translate them into physical realities 【9】. The technical specifications for which hardware will be acquired and which programs will be written are developed as part of the design step 【10】.

Programmers

Computer programmers write instructions to process data and produce information. Using programming languages, programmers develop the applications programs that operate the computer hardware.

The task of programming involves much more than the simple coding of instructions in a computer language. Programming includes determining the computer processing functions to be performed, arranging those functions in an organized fashions, selecting the specific computer operations that will implement the functions, coding the operations in a programming language, and testing the program on a computer 【11】.

Common jobs for programmers are as applications programmers, who write programs to perform business functions identified by the systems analyst and designed by the systems designer, and maintenance programmers. Maintenance programmers revise or modify applications programs in accordance with processing changes that inevitably arise when a system is in operation for a long period of time 【12】.

Computer Operators

Computers are operated at three general levels within organizations. At the individual work station level, individual users own and/or operate their own personal computers. They purchase software from computer stores. These individuals seldom write their own programs instead, they rely on commercial software packages. Programs typically are purchased to perform word processing, financial analysis, electronic filing and data management, and communications with other computers. At the information center level, a centralized facility or a satellite center within an organization is maintained.Users work in the information center to perform their own data processing. The information center usually is equipped with personal computers, computer terminals tied into the central computer, and other user-accessible equipment. This hardware is supported by a wide range of software designed for the nontechnical user. Rather than make formal requests for processing from the computer service department, people can go directly to the information center to satisfy their processing needs.

At the information service department level, a formal production center is used for complex information processing. In many organizations, especially in medium-and large-size companies, specialist are required for operating computers. Running a large computer services department is, in itself, a complex, manufacturing-type operation. Within these production environments, many types of people, with many levels of responsibility, are required:

(1) Data entry operators transcribe data from source documents to input media for entry into the computer system.

(2) Console operators monitor the system through terminals attached to the processor to make sure that the system is running properly and to recover system operations from any problem.

(3) Peripheral device operators assist in the movement of work through the computer center. These operators operate the all kinds of equipment that are involved in processing.

(4) Production schedulers schedule, control, and monitor the flow of work through the computer system.

(5) Data librarians maintain the inventory of programs, operating manuals, and data storage media to make sure that these resources are accounted for and available when needed.

Technical Specialists

Within larger computer installations, people will be assigned to make sure that existing systems continue to run smoothly. Included in this category are several technical specialists:

(1) Systems programmers maintain the operating system and other system software to ensure that the computer is providing needed services to the systems development specialists, the applications programmers, the operations staff, and the end users.Systems programmers maintain the system software purchased with the computer and write special routines to facilitate use of the system.

(2) Field engineers, who represent the computer manufacture, diagnose, repair, and maintain computer hardware.

(3) Telecommunications specialists serve as consultants to computer systems that are tied into companywide, nationwide, or worldwide communications networks.These specialists have an expertise in the integration of computers and communications technology.

(4) Database administrators offer specialized knowledge of methods of organizing the massive data resources that can exist within a company's storage files. These organization methods facilitate the maintenance and retrieval of data resources.

5.2.2 Specialized English Words

computer software　计算机软件
perform on data　对数据进行处理
business programming languages　商务程序设计语言
hardware circuits and devices　硬件电路和设备
systems analysts　系统分析人员
functional specification　功能约定
blueprint　（工程中运用的）蓝图
programming language　编程语言
commercial software package　商业化的软件包
financial analysis　财务分析
data management　数据管理
information center　信息中心
central computer　中央计算机
data entry operator　数据输入操作员
input media　输入媒介
operating manual　操作手册
technical specialist　技术专业人员
operating system　操作系统
system development specialist　系统开发专业人员
operation staff　操作人员
field engineer　现场工程师
companywide, nationwide, or worldwide communication network　公司范围、全国范围、

或是世界范围内的通信网络

company storage file   公司的存储文件
computer hardware   计算机硬件
machine-language   机器语言
general-purpose programming languages   普通程序设计语言
storage space within memory   主存储器中的存储空间
information processing professionals   信息处理专业人员
physical reality   具体实体
technical specification   技术规约
individual work station   独立工作站
word processing   文字处理
electronic filing   电子文档
computer terminal   计算机终端
personal computer   个人计算机
nontechnical user   非专业用户
source document   源文件
inventory of program   程序清单
data storage media   数据存储媒介
systems programmer   系统程序员
system software   系统软件
applications programmer   应用程序员
end user   终端用户
telecommunication specialist   远程通信专业人员
database administrator   数据库管理人员
specialized knowledge   专业知识
massive data resource   巨型数据资源

5.2.3 Notes

【1】So, in fact, the user's program only specifies what operations are to be performed, not how to perform them.

应该看出，该句主句的谓语动词有两个宾语，一个是由 what 引导的宾语从句 what operations are to be performed，另一个是由 how 引导的不定式短语 how to perform them。因此，全句应译为："因此实际上,用户的程序只能指定计算机执行何种操作,而不能指导它如何操作"。

【2】Programs that act upon instructions provided in application programs are known collectively as system software.

句中，collectively 一词为"总体上"之意，此处译为"统称"。因此，全句应译为："根据应用程序所提供的指令而执行动作的程序统称为系统软件"。

【3】A computer cannot perform usually unless there is a user group that has a need for the results of processing.

句中 unless 的意思是"除非……，否则……"，而主句又是否定式，若直译译文会很别扭，因此将 unless there is 译为"如果没有"。全句应译为："如果没有对处理结果有需求的用户群，计算机通常是不可能有所表现的"。

【4】Although they may not have a "hands-on" need for computers, users play a critical role in defining the tasks computers perform and the products they deliver.

句中，动名词 defining 有两个并列的宾语 the tasks 和 the products，只不过这两个宾语又分别由（that）computers perform 和（that）they deliver 两个定语从句来修饰。因此全句应译为："尽管用户不可包揽对计算机的所有需求，但是其在设定计算机所需完成的任务，以及其所需提供的结果这两方面起着至关重要的作用"。

【5】On the other hand, computer professionals must understand that if they did not produce useful products, they wouldn't be needed.

全句译为："另一方面，计算机专业人员必须懂得，如果他们未生产有用的产品，他们就不会被需要"。

【6】Together, users and analysts study the business environment and develop an understanding of the business procedures to be computerized, the data to be processed, and information to be delivered.

虽然该句也有这么长，但其实是一个简单句，句中两个谓语动词 study、develop 是并列的，因此全句应译为："用户与分析师一道共同研究商务环境，从而得出对用于计算机的商务处理程序、需处理的数据和需提供的信息的深刻理解"。

【7】Then, these functional specifications become the blueprints for designing a computer-based system.

句中的 specifications 是"详细叙述"的意思，此处转译为"具体约定"。因此全句应译为："这样，这些对功能的具体约定便成了设计计算机应用系统的蓝图"。

【8】Systems designers work from these documents to design the actual configurations of hardware, software, data files, manual procedures, communication networks, and human resource that will be required to implement the system.

此句的句干为 Systems designers work to design the actual configurations，而介词短语 of hardware, software, data files, manual procedures, communications networks, and human resource 是宾语 the actual configurations 的后置定语，说明是哪些东西的"实际框架"。而由 that 引导的定语从句 that will be required to implement the system 则是修饰其前面整个介词短语 of hardware, software, data files, manual procedures, communications networks, and human resource 的，因此全句应译为："系统设计师根据这些文件设计出硬件、软件、数据文件、手工操作步骤、通信网络和人力资源等，这样一些完成计算机系统设计的必要条件的实际框架"。

【9】Designers take the functional requirements identified during the analysis step and translate them into physical realities.

该句中的两个谓语动词 take 和 translate 是并列的，表示两个几乎同时并有递进关系的动作，因此全句应译为："设计人员掌握这些在分析阶段所确立的对计算机系统的功能要求，并将其转化为具体的实体"。

【10】The technical specifications for which hardware will be acquired and which programs will be written are developed as part of the design step.

该句的句干为 The technical specifications are developed as part of the design step，其意为

"这些技术规约会演变成设计步骤的一部分"。而介词 for 引导的介词短语 for which hardware will be acquired and which programs will be written 做主句的原因状语。只不过该介词短语中，介词 for 的宾语是两个并列的宾语从句：which hardware will be acquired 和 which programs will be written。因此全句应译为："这些技术规约会演变成设计步骤的一部分，因为其与计算机硬件的选定和程序的编写紧密相关"。

【11】Programming includes determining the computer processing functions to be performed, arranging those functions in an organized fashions, selecting the specific computer operations that will implement the functions, coding the operations in a programming language, and testing the program on a computer.

该句中的谓语动词 includes 有五个宾语，其均为动名词短语：determining the computer processing functions to be performed, arranging those functions in an organized fashions, selecting the specific computer operations that will implement the functions, coding the operations in a programming language 和 testing the program on a computer。只不过第三个短语中又有一个从句就是了。因此全句应译为："程序编辑包括确定需执行的计算机处理功能，以一种有序的方式安排这些功能，选择计算机实现这些功能的具体操作方法，用程序语言对操作指令进行编码，以及用计算机测试这个程序"。

【12】Common jobs for programmers are as applications programmers, who write programs to perform business functions identified by the systems analyst and designed by the systems designer, and maintenance programmers. Maintenance programmers revise or modify applications programs in accordance with processing changes that inevitably arise when a system is in operation for a long period of time.

这段话虽然有两句，但由于其间的联系较为紧密，因此归并在一起翻译。当然，这样一来其内容就较多，而汉语又以短句表示，故应该综合起来译成几句。因此全句应译为："计算机程序员的工作通常有两种，一种是应用程序员的工作，另一种是维护程序员的工作。应用程序员编写程序，以执行由系统分析师所指定和系统设计师所设计的商务功能。维护程序员校正应用程序，或根据处理过程的变化（这在系统运行了较长一段时间后是难免的）修改应用程序"。

5.2.4 Translation

计算机软件及人力资源

通过计算机软件或程序的应用，计算机的硬件才得以为人们工作、娱乐及日常生活服务。

程序语言

程序语言是一系列具体的和具有执行步骤的一系列指令，其在计算机进行数据处理时必定要用到。这些指令用计算机能识别的专用程序语言编写。现在已经产生了 200 多种不同程序语言。然而，仅仅只有其中的 10 来种才是常用的。

多数计算机程序采用类似于标准的英语和/或数学术语写成。因此，这些程序能写成人类可读的形式。这种程序所具有的这些特性，有助于人们用自己的语言，而不是用计算机的二进制语言表述问题。而后，计算机用语言翻译程序将这些类似于英语的指令，转换成能由计算机用来执行指令的机器语言。

程序语言的种类

（1）商务程序语言被设计成解决商务问题。商务问题常常要求存储和处理大量数据。通常商务处理需要一些对数据的基本算术运算。商务程序语言是专门为方便及有效地满足处理需要而产生的。像 COBOL（公用商务语言）和 RPG 语言（报告生成程序）便是在商务信息处理时广泛使用的两种语言。

（2）科学程序语言是一种用于数据处理量较小，而具有相对复杂数学处理过程的语言。科学的、工程的、数学的应用需要进行专门的计算，而这类语言便自动提供了这些计算。像 FORTRAN 语言就是一种广泛应用的科学程序语言。

（3）普通程序语言是这样设计生成的，其将商务语言数据处理及表格生成能力，与科学程序语言的数学处理功能相结合而形成的一种程序语言。像 BASIC 语言、PL/1 语言和 Pascal 语言等这样一些普通程序语言，几乎可有效地运用于任何类型处理过程。普通程序语言常用于教学生如何进行编程。

除上述这些语言外，程序员还可以利用其他语言处理文本、绘制图形和完成其他专门应用。

计算机软件类型

计算机软件可划分为两大范畴：
（1）应用软件。
（2）系统软件。

应用软件

应用软件用诸如 COBOL、PRG、BASIC 和 FORTRAN 等语言编写程序，专门用于满足商务和/或科学处理的需要。这些语言允许人们用像英语一样的形式表达它们的处理要求。

计算机不能直接执行应用程序指令。计算机仅能理解用 0 和 1 序列编写的二进制语言。因此实际上，用户的程序只能指定计算机执行何种操作，而不能指导它如何操作。因此，计算机需要能确切地告诉硬件如何去执行所要求的操作的指令。

系统软件

根据应用程序所提供的指令而执行动作的程序统称为系统软件。这些程序解释包含在应用程序中的用户程序指令，并驱动相关的硬件电路或装置去完成程序的处理。系统软件是与计算机硬件一并销售的。根据计算机操作系统（OS）的不同，也有的计算机软件系统打包装在一个磁盘内，其称为磁盘操作系统（DOS），使用时可将其存入计算机的存储器。在某些计算机中，特别在家用和个人计算机中，操作系统或其中的一些部分已永久装入计算机存储电路中。

因此，系统软件是用户应用程序和计算机硬件的中间媒体。例如，当用户程序进行输入操作时，程序系统软件便调出某个系统程序，去驱动输入装置，接收数据并将其送入存储器中，并在存储器中分配存储空间以便存储数据。

正如你已注意到的那样，有好几种计算机指令可供使用。作为一个用户，可通过应用程序向计算机提供要处理的数据这种方式与计算机进行交流。数据锁存处理功能就置于该程序内。然后，反过来，应用软件又会对系统软件发出需要进行具体操作的指令。因此，系统软件与计算机硬件相互交流，一起完成实际的操作过程。由于有系统软件的缓冲作用，作为一个用户，可免于考虑那些复杂高深的技术问题。

人力资源

不管计算机变得如何巨大或如何复杂,其中人仍然起着关键的作用。计算机应用范围是如此之广,以至于牵涉到方方面面的人,他们都扮演着不同的特殊角色,包括:

(1)用户,即客户,其确立自己对计算机的需求,并极大限度地运用计算机为其提供的处理结果。

(2)系统分析和设计员,其分析用户的需求情况,并根据其采用的计算机设计出相应的系统,以满足用户的需要。

(3)程序员,其设计、编写和维护可满足用户需要的应用软件。

(4)计算机操作员,其在计算机工作室操作计算机,以便处理数据和向用户提供信息。

(5)技术专家,其维护计算机硬件、系统软件及数据资源。

以上(还有其他一些)专家汇集于专门从事计算机工作的机构,向那些有计算机服务需要的人们提供帮助。

用户(客户)

计算机系统需要用户。这些用户有业务需求,并且指定要用计算机来提供专门的服务。如果没有对处理结果有需求的用户群,计算机通常是不可能有所表现的。尽管用户不可包揽对计算机的所有需求,但是其在设定计算机需要完成的任务,以及它需要提供的结果这两方面起着至关重要的作用。

因为这种关键性作用,所以用户与信息处理专业人员之间的联系是很重要的。用户必须能与计算机专业人员进行交流,以便指出他们的需求,为什么需要,什么时间任务必须完成。另一方面,计算机专业人员必须知道他们的产品是否有用,他们是否不被用户所需要。

如果欲使计算机能成功地应用并获利,相互理解和合作就是警句名言。

系统分析师和设计师

在任何大型公司中,都能找到这样一些技术专家,他们与用户合作为其研制计算机的应用系统,从而让计算机为用户工作并得出其所需要的结果。在系统研究领域,其最常见的技术头衔就是系统分析师。这些技术专家分析商业需求,并研制计算机应用系统便满足这些需求。正是在这个过程中,计算机的系统分析师和设计师在计算机用户的日常商务领域和计算机专业技术人员的高科技世界之间形成了一个相互连接的桥梁。

系统分析师有时单独工作,有时也与用户组成项目组一起工作。用户与分析师一起共同研究商务环境,从而得出对用于计算机的商务处理程序、需处理的数据和需提供的信息的深刻理解。用户和分析师一起建立一套文件,以确定哪些是计算机应用系统应具备的功能。这样,这些对功能的具体约定便成了设计计算机应用系统的蓝图。

系统设计师根据这些文件设计出硬件、软件、数据文件、手工操作步骤、通信网络和人力资源等,这样一些完成计算机系统设计的必要条件的实际框架。设计人员要掌握这些在分析阶段所确立的对计算机系统的功能要求,并将其转化为具体的实体。这些技术规约会演变成设计步骤的一部分,因为其与计算机硬件的选定和程序的编写紧密相关。

程序员

计算机程序员编写处理数据和产生信息的指令。运用编程语言,程序员编辑出操作计算机硬件的应用程序。

程序编辑所涉及的任务,要远远多于单单是对计算机语言中的指令进行编码。程序编辑包括决定需执行的计算机处理功能,以一种有序的方式安排这些功能,选择计算机实现这些功能的具体操作方法，用程序语言对操作指令进行编码，以及用计算机测试这个程序。

计算机程序员的工作通常有两种，一种是应用程序员的工作，另一种是维护程序员的工作。应用程序员编写程序，以执行由系统分析师所指定和系统设计师所设计的商务功能。维护程序员校正应用程序，或根据处理过程的变化（这在系统运行了较长一段时间后是在所难免）修改应用程序。

计算机操作员

在计算机的所有用户机构中，其对计算机的操作运用分为三个层次。处于独立工作站层面上的用户拥有或操作他们自己的计算机。他们从计算机商店购买软件。这些用户很少编写自己的程序，而是依赖于商业化的软件包。这些商业化的程序主要用于完成文字处理、进行财务分析、建立电子文档、执行数据管理及与其他计算机进行通信联系。处于信息中心层面上的用户，其单位内部就拥有计算机中心。他们在信息中心进行自己的数据处理。信息中心通常配有多台个人计算机，其终端均与中央计算机及用户的其他装置相连。整个硬件由一个专门为非专业用户而设计的大规模软件支持。如有工作需求，这种用户不需要向计算机服务的专业部门提出正式请求，而只要直接进入本单位的信息中心便可满足自己的工作要求。

处于信息服务中心机构层面上的用户，其为了完成复杂的信息处理过程，在其单位内部设立了一个正规的生产中心。在许多信息服务中心机构中，尤其是那些中大型公司，要求有专业人员操作计算机。一个大规模计算机服务公司运转的本身，就是一个复杂的生产型的运作过程。在这些生产环境中，需要许多不同类型，负有不同职责的人员。

（1）数据输入操作员将源文件中数据转换成人与计算机之间的输入媒介，以便输入计算机系统。

（2）控制台操作员通过连接到中央处理器的终端监控整个系统，以便了解系统是否运行正常，且在出现故障时恢复正常运行。

（3）外围设备操作员通过计算机中央协助计算机运行。他们操作在处理过程中将会运用到的各种外设装置。

（4）生产计划员通过计算机系统计划、指挥和监控工作流程。

（5）数据库管理人员维护各种程序、操作指令和数据存储媒介的清单，以便确保这些资源不丢失，在需要时可调出使用。

专业技术人员

在大型计算机站，会有专职人员日夜值班确保当值系统能连续不断地顺利运行。这其中的专业技术人员有如下几种。

（1）系统程序员，其维护操作系统和其他系统软件，以确保计算机为系统研究专业人员、应用程序员、各种操作人员和终端用户提供必要服务。系统程序员维护随计算机一同购买的系统软件，编写专门的引导程序以便于系统的使用。

（2）现场工程师，其代表计算机制造商在工作现场负责对计算机的诊断、修理和维护硬件的工作。

（3）远程通信专业技术人员，其为接入或是公司范围，或是全国范围，或是世界范围内的通信网络的计算机系统，提供专业技术咨询。这些专业人员具有计算机与通信技术综合运用的专长与经验。

（4）数据库管理人员，其具备对存在于公司存储文件中的巨型数据资源进行组织和管理方法的专业知识，该组织和管理方法便于数据资源的维护和检索。

Unit 5.3   Personal Computer Word Processing and Electronic Spreadsheets

5.3.1   Text

Software, or programs, for the personal computer usually is purchased from software vendors. These software packages provide a broad range of services for people in operational and managerial-level positions. Typical software packages include:

(1) Word processing
(2) Electronic spreadsheets
(3) Data management
(4) Business graphics
(5) Data communication

Word Processing

Word processing refers to the use of computers to handle text and to produce typed and/or printed documents. Text in the form of memos, letters, and reports is keyed into the computer, retained in memory, manipulated electronically, and printed and/or stored on disk. Because text can be viewed on a video display screen during writing and editing, any insertions, deletions, and other editing changes in the text are made easily. Perfect copy can be produced before a single character is printed 【1】. After the document is completed and proofread, it can be routed to the printer and/or stored to disk for filing and later retrieval and printing.

A variety of word processing programs is available commercially【2】. Some typical features of these programs are described in the sections below.

(1) Electronic text capture. As text is entered on the keyboard, the program stores the characters in memory and displays, or echoes, them on the video display screen. The cursor, an underline or black character appearing on the screen, indicates the current typing position on the line. If a word contains more characters than can fit within the right margin, the entire word automatically is "wrapped around" in the screen and placed at the beginning of the next line. This word wrap feature allows the user to enter text continuously, uninterrupted by either line feeds or carriage returns.

(2) Full-screen editing. Text insertions, deletions, and changes can be made easily to text displayed on the screen. To insert new text within a document, the cursor is moved to the position at which a new word, sentence, or paragraph is to be added. By depressing an "insert" key, space is created between the existing text to allow typing of the new material. Following the insertion, the text is closed up, with the computer adjusting line lengths and spacing to integrate the additional text【3】. To delete text, the cursor is moved to the beginning of the word, sentence, or paragraph

that will be removed. A "delete" key is used to mark or highlight the unwanted text and to erase the text from the document. Again, spacing is adjusted automatically to close the surrounding text. Other changes or error corrections are made by typing the replacement text over the incorrect or unwanted text.

(3) Move and copy. Words, sentences, paragraphs, and entire pages of text can be moved or copied anywhere within the document. The block of text to be moved or copied is marked and the cursor is placed at the location in the document where the text is to appear【4】. Then, the appropriate keystrokes are entered to instruct the program to place the text at the new location. The program automatically adjusts the surrounding text to make room for the new block of text. Many word processing programs can copy portions of a file or even an entire file from a disk to a document file and copied as needed into new documents.

(4) Search and replacement. The "search and replace" feature allows for selected changes to be made to some or all occurrences of a particular character, word, or phrase【5】. The computer searches through the document, character by character, to find the indicated text【6】. Depending upon the instruction the operator provides, the computer either stops at each indicated occurrence and allows the operator to change the wording, or automatically replaces the text with the requested change【7】.

(5) Column tabulation. Within many word processing programs, automatic tabulation is available to align columns of text or numbers【8】. A special feature of many word processing packages is the automatic alignment of numbers around decimal points, and the capability for row and column arithmetic【9】. For example, a table of numbers appearing in the text can have row and column totals generated by the software to provide checks on arithmetic.

(6) Headers and footers. Blocks of text, such as page heading, that will appear on every page of the document can be entered once and then copied automatically onto each subsequent page【10】. This feature is commonly used to identify document pages and to provide automatic page numbering. Some word processors can keep track of footnote references and place them on the proper pages of the document.

(7) Spell check. Many word processing packages are supported with electronic dictionaries and processing routines to allow automatic checks for spelling errors. These spelling checkers advance through the text and compare each word with the word list stored in the dictionary【11】. When differences are found, the computer highlights the unrecognized word and allows the operator to correct it. If the word is correct but does not appear in the dictionary, the operator can add it to the word list.

(8) Mail-merge. List of names and addresses can be stored separately from a document【12】. During printing, these names and addresses can be merged with documents to personalize form letters and memos. Data, mailing addresses, salutation lines, and other individualized blocks of text can be integrated within documents so that a standard document does not have to be retyped for each addressee【13】.

(9) Print format. Documents are printed according to formats selected by the operator. Options typically permit choices of type styles and pitches, automatic page numbering, right-margin

justification, and printing of multiple copies【14】. Editing a document can change the page lengths that were set during original text entry. Therefore, word processors often have repagination features that automatically readjust page lengths and reformat the text to meet standards required for printing.

The above features are common to professional word processing software packages-these that will be used in a business or professional office for text processing. Some packages have more elaborate features than the ones described above and others have less. The wide variety of available word processors allows the user to choose the package that is most compatible with his or her text creation, editing, and printing needs.

Electronic Spreadsheets

Each cell within the spreadsheet can contain one of three kinds of entries:
(1) A heading identifies the data contained in a particular row, column, or cell.
(2) A numeric value represents the financial or statistical data to be processed.
(3) A formula indicates the calculation necessary to derive a value for a particular cell.

Assume, for example, that a company plans to use a spreadsheet to analyze sales revenues and production costs for a six-month period. In setting up the worksheet, the first row and the first column of the table would be used to enter the heading information. This information identifies the row and column values. Then, the actual revenue and expense amounts would be keyed into six consecutive columns, with each column representing a one-month period【15】.

When these original data have been placed in the spreadsheet, subsequent calculations can be carried out automatically. For instance, formulas can be supplied to provide a total for revenues and a total for expenses for each month. After these formulas have been entered into the worksheet, the calculations can be performed by the computer, which outputs a comparision of the two amounts over the six-month period【16】.

As a further use of the spreadsheet, the actual amounts could be used to project future revenues and expenses. Formular relating revenues to expenses can be entered. Then, the computer automatically would calculate expected figures for any number of future monthly periods. The projections would be placed in additional columns of the worksheet.

Spreadsheet packages offer a variety of features to assist in financial planning, budgeting, and forecasting. The following features are common to many electronic spreadsheets.

(1) Automatic calculation. After a variety have been loaded into the worksheet, calculations are carried out automatically by the computer. Both rows and columns can be extended to include results of computations. As another option, the value that will appear in a particular cell can be defined by a formula. The formula indicates the mathematical operations to be performed on values contained within specified cells. The formula also indicates the order in which computation are to be performed and the location of the cell being defined.Then, the value that results from the computation is placed in the designated cell.

The intersections of rows and columns represent cells into which data are placed.The width of the cells can be varied to match the size of data items.For example, headings may be allowed to spill over into adjacent cells, or a single cell may be defined to hold a complete page of text.The sizes of cells may be reduced to hold smaller-than-standard values so that more of the spreadsheet can be viewed on the screen.

(2) Templates. Templates, or standard formulas, are provided with many spreadsheet packages. These templates can be used to set up the worksheet to perform typical data processing tasks. For example, standard templates may be provided for analyzing revenues and expenses, projecting cash flows, establishing budgets, forecasting trend, preparing financial statements, and several other common business functions.Availability of these templates, or financial models, means that the user does not have to build special tables each time the spreadsheet program is run.The selected template can be called up to format the worksheet and to allow entering of data under standard formats.

(3) Move and copy. Spreadsheets are designed for ease of expansion and contraction. Individual cells or ranges of cells can be moved or copied from one location in the worksheet to another. Great flexibility is provided to rearrange rows and columns for convenience of analysis or for presentation impact 【17】. The software adjusts formulas automatically to take into account the new locations of referenced cells.

(4) Add and delete. It is also quite easy to add new rows and columns to the spreadsheet or to erase ranges of cells. These features assure that the worksheet remains dynamic. As in the case of the move and copy facilities, formulas and cell relationships are maintained automatically by the software whenever row and column references are changed 【18】.

(5) Scrolling and windowing. Spreadsheets often become larger than the video display screen, especially if 25 or more rows and 10 or more columns are used.In these cases, the package allow horizontal and vertical scrolling to bring a particular section of the worksheet onto the screen.In other cases, windowing is used.This technique allows two or more small parts of the worksheet to be viewed simultaneously within separate windows, or sections, of the screen.

(6) Report writing. Screen displays can be routed to a printer for hard-copy output. The user can print selected portions of the worksheet or the entire spreadsheet.Most packages have options for copying parts of the worksheet into word processing programs so that tables can appear within reports.

(7) Graphics. The values that appear within spreadsheet cells can be displayed graphically as charts and graphs, as the special processing routines convert figures into pictures with full labeling. The size of the chart or graph can be modified to fit onto the video display screen or paper. On some displays and printers, the output can be viewed in various colors for emphasis and attractiveness.

(8) File loading. Values can be placed within the cells of a spreadsheet in one of two ways: The user can key the figures into the table using simple editing commands to insert, delete, change, and format values, or the values can be loaded into the table from a data file 【19】. For example, financial information that is kept current within company files can be selected and entered into the table automatically, avoiding the need to re-key the data each time they are analyzed.

Electronic spreadsheets allow the user to perform a wide range of calculations and to create displays automatically. For the most part, spreadsheets are used to analyze the current financial status of a business and to project the likely results of planned operations. For example, the user can prepare a spreadsheet containing current financial data related to a business situation. Then, selective changes can be made to the data under various assumptions about the impacts of certain plans, polices, and trends. These assumptions are provided as formulas relating two or more of the cells of data in the worksheet. The spreadsheet program, then, makes the necessary calculations to project future business results. The business manager can view the possible results of plans and policies before they are implemented.

5.3.2 Specialized English Words

software package   软件包
word processing program   文字处理程序
electronic spreadsheet   电子表格
business graphics   商务绘图
full-screen editing   全屏幕编辑
search and replacement   搜索及替换
column tabulation   竖式制表
line feed   换行
search and replace feature   搜索及替换功能
align column   列对齐
row and column arithmetic   行和列的算术运算
page heading   页首
electronic dictionary   电子词典
personalize form letters and memos   私人性质形式的信件和备忘录
spreadsheet program   电子表格生成程序
scrolling and windowing   滚动及多窗口显示
operational and managerial-level   操作和管理层面
data management   数据管理
data communication   数据通信
electronic text capture   电子文本输入
move and copy   移动及复制
available commercially   在市面上可买到
wrap around   另换一行
carriage return   回车
selected change   选择性地修改
around decimal point   围绕小数点，在小数点的前后
blocks of text   文本块
footnote reference   下角标注
spelling checker   拼写检查器
mail-merge   通信格式生成

individualized blocks　　各个不同的子块
sale revenue　　销售收入
production cost　　生产成本
full labeling　　全标识
file loading　　文档下载

5.3.3　Notes

【1】Perfect copy can be produced before a single character is printed.

　　句中 perfect copy 意为"完美无缺的文本"，既然完美无缺的文本可在第一个文字打印出之前生成，岂不就是说"待完美无缺的文本生成后再打印"吗？因此该句应译为："（这样）所生成的文本可在处理得完美无缺后再进行打印。"

【2】A variety of word processing programs is available commercially.

　　句中 available commercially 意为"在商业行为中可得"，而如若直译为"多种文字处理程序可在商业行为中得到"不合汉语的习惯，因此转译为："在市面上可买到多种（版本）的文字处理程序。"

【3】Following the insertion, the text is closed up, with the computer adjusting line lengths and spacing to integrate the additional text.

　　句中 following the insertion 意为"紧跟插入过程"，此处译为："插入过程完成后"。而 with the computer adjusting line lengths and spacing to integrate the additional text 是由 with 引出的介词短语，做句子 the text is closed up 的状语，表示一种伴随状态。因此该句应译为："插入过程完成后，随着计算机调整每行的长度和安置欲插入的文本的工作的进行，新的文本便合成了。"

【4】The block of text to be moved or copied is marked and the cursor is placed at the location in the document where the text is to appear.

　　该句的整体构架是一个由并列连词 and 连接而成的并列复合句，而连词 and 的含意中又有明显的时间上的前后次序，以及动作上先后递进的关系。因此该句应译为："先将欲移动或复制的那整块文本做上记号，然后将光标放置在该块文本在文件中所出现的地方。"

【5】The "search and replace" feature allows for selected changes to be made to some or all occurrences of a particular character, word, or phrase.

　　句中 allows 若直译为"允许"会很别扭，改译为"可实现"更好。因此该句应译为："搜索及替换功能可实现对选定的字母、单词或短语的某些部分或整体进行有选择性的修改。"

【6】The computer searches through the document, character by character, to find the indicated text.

　　在该句中应注意将 through the document 的意思翻译出来，否则其译意便没到位。through the document 的含意为"通篇，贯穿整个文件"。因此该句应译为："计算机在整个文件中逐字逐字地进行搜索，以便查找出指定的修改文件。"

【7】Depending upon the instruction the operator provides, the computer either stops at each indicated occurrence and allows the operator to change the wording, or automatically replaces the text with the requested change.

　　句中的 Depending upon the instruction the operator provides 要译为："根据操作员所提供的

不同指令"。此外还应注意该句的句干是 the computer either stops..., or automatically replaces..., 其实是一个由并列连词 either ..., or...引出的并列复合句。只是其第一个并列子句中的主语有两个并列的谓语动词 stops 和 allows 就是了。因此该句应译为:"根据操作员所提供的不同指令,计算机或是停止在每一指定的更改处,让操作员对字符进行更改,或是用所需的更改内容自动替换原文本。"

【8】Within many word processing programs, automatic tabulation is available to align columns of text or numbers.

句中 to align columns 意为"列对齐", is available 意为"可找得到的"。故该句可译为:"在许多文字处理程序中,可运用自动列对齐程序对文本或数据进行列对齐。"

【9】A special feature of many word processing packages is the automatic alignment of numbers around decimal points, and the capability for row and column arithmetic.

句中 around decimal point 意为"围绕小数点","在小数点的前后",row and column arithmetic 意为"行和列的算术运算"。故该句应译为:"许多文字处理软件包的特定的功能,就是围绕小数点对数字自动进行列对齐,并能进行行和列的算术运算。"

【10】Blocks of text, such as page heading, that will appear on every page of the document can be entered once and then copied automatically onto each subsequent page.

句中 blocks of text 意为"文本板块",page heading 意为"页首"。故该句应译为:"像文件的每一页的开头都会出现的(具有固定格式的)页首那样的文本块,一旦一次性输入后,便可自动地复制到后续的各页中。"

【11】These spelling checkers advance through the text and compare each word with the word list stored in the dictionary.

在该句的翻译中应注意 through 这个词的含义,其意为"在其中穿过",千万不要译成汉语的"通过……方法"。因而 through the text 的意思便是"通篇的,贯穿整篇的"。故该句应译为:"这些拼写检查器将整个文本中的字词,逐个地与词典中存储的字词进行比较。"

【12】List of names and addresses can be stored separately from a document.

该句看似简单,但有几处要注意的地方,否则就会译出差错。首先,list 在此处不是表格的意思,而是"清单"的意思。此外,对于 stored separately from a document,要注意其中的 from 的含意,其意为"与……分开"。因而该句应译为:"经常保持通信联系的人的名单和地址,可与文件分开存放。"其中,"经常保持着通信联系的人的"这几个字是运用增译法加上去的。

【13】Data, mailing addresses, salutation lines, and other individualized blocks of text can be integrated within documents so that a standard document does not have to be retyped for each addressee.

句中 a standard document does not have to be retyped for each addressee 是由从属连词 so that 引导出的结果状语从句,因此应在从句前译出"这样"。故该句应译为:"数据、通信地址、称呼及文本中其他各个不同的子块都能与文件拼接组合形成整体,这样对每个通信对象来说,其标准文件就不必每次都重打印了。"

【14】Options typically permit choices of type styles and pitches, automatic page numbering, right-margin justification, and printing of multiple copies.

句中 options 意为"选项",typically 不要翻译成"典型的"(不合中国人的习惯),而应翻译成"通常的"。此外,句中所列选项较多,故翻译时运用增译法添加"有如下几种"几个字。因此该句应译为:"通常的选项有如下几种:字体和行距、页码自动标记、右边界对齐,以及多份打印等。"

【15】Then,the actual revenue and expense amounts would be keyed into six consecutive columns,with each column representing a one-month period.

句中 with each column representing a one-month period 由介词 with 引出的为一介词,作句子的状语,表示一种伴随状态。因此该句译为:"然后,将实际的收入及支出之值输入到六个相邻的列中,每一列代表一个月的时间周期。"

【16】After these formulas have been entered into the worksheet, the calculations can be performed by the computer, which outputs a comparision of the two amounts over the six-month period.

句中 which outputs a comparision of the two amounts over the six-month period 是由从属连词 which 引出的非限定性定语从句,表示一种伴随的结果,对主句动作的结果做进一步地说明。因此该句译为:"当该运算公式进入表格后,其通过计算机运算的结果,便是六个月内的总收入和总支出的比较情况"。

【17】Great flexibility is provided to rearrange rows and columns for convenience of analysis or for presentation impact.

首先该句为被动语态不便直译,应转译为主动语态。此外,句中的 presentation impact 直译为"呈现的印象",转译为"视觉效果"。因此该句译为:"为了分析的方便和整个表格的视觉效果,软件在调整行和列的方面具有很大的灵活性"。

【18】As in the case of the move and copy facilities, formulas and cell relationships are maintained automatically by the software whenever row and column references are changed.

句中 whenever 所引导的 whenever row and column references are changed 为一时间状语从句,表示主句的伴随时间。whenever 的意思是"无论什么时候",但按中国人的习惯译为"一旦"会更好些。此外,句中的 row and column reference 意为"行和列的标记",此处译为"行和列的计数",相信读者更易理解。因此该句译为:"当处于和移动及复制状态时,一旦行和列的计数改变,软件会将运算公式与其所处的方格之间的关系自动地维持下来"。

【19】Values can be placed within the cells of a spreadsheet in one of two ways: The user can key the figures into the table using simple editing commands to insert, delete, change, and format values, or the values can be loaded into the table from a data file.

句中的这两句话联系得比较紧密,且前一句中出现的 in one of two ways 其意(两种方法中的一种)在后一句中并没有承接之处,而是用了…,or…的句式,如若直译就有前后不相衔接的感觉。故应将此两句拆开,重新安排译文。这样该句便译为:"数据可以用以下两种方法放置于表格中的方格内:一种是用户运用简单的编辑命令,将数据以插入、删除、修改或格式化等方法输入表格中。另一种是将数据从数据文档中下载到表格内"。

5.3.4 Translation

个人计算机的文字处理和表格生成

个人计算机通常使用的软件或程序是从软件商那里购买的。这些软件包为人们在操作和管理这个层面上的工作提供一个广阔的服务范围。典型的软件包有以下几种:

·187·

（1）文字处理
（2）电子表格
（3）数据管理
（4）商务绘图
（5）数据通信

文字处理

　　文字处理是指运用计算机处理文本及生成打印文件。以便条、信件、报告等形式出现的文本由键盘输入计算机，暂存在存储器中，经过电子处理，打印出来或存储在磁盘中。因为文本可在写入和编辑过程中在显示屏上看到，因此像插入、删除及其他一些编辑方面的更改便可很容易地进行。这样，所生成的文本可在处理得完美无缺后再进行打印。当文件完成并校对后，便可输入打印机打印，或者存入磁盘存档以便以后检索和打印。

　　在市面上可买到多种（版本）的文字处理程序。这些程序所具有的一些典型特点说明如下。

　　（1）电子文本输入　　当文本从键盘输入时，程序就将文字暂存在存储器中，并将其在显示屏上显示。光标（一种出现在显示屏上线状或块状字符）所表示的位置就是所在的这一行当前的输入位置。如果一个词中所包含的字符超出右边的空间所能容纳的范围，则整个词便会自动另换一行下移至下一行的行首。这种词的换行功能使用户可连续不间断地输入文本，而不被换行或回车所打断。

　　（2）全屏幕编辑　　当文本显示在显示屏上时，便能很方便地进行插入、删除和更改等编辑工作。若要在一个文件中插入新的文本，只要将光标移到那些词、句子或者段落等欲插入的地方。通过按一下"插入"键，便在原文本中产生出空间，这样便可输入新的内容。插入过程完成后，随着计算机调整每行的长度和安置欲插入的文本的工作的进行，新的文本便合成了。若要删除文本，只要将光标移到欲删除的字词、句子或段落的起始位置。按下删除键，以便在不要的文字内容下作出标记，从而将其从文件中删除。其后，整句的空间会再次自动调整，以靠近周围的文本内容。另一种变更或修改错误的方法，是将所需的文本内容覆盖在错误的或不需要的文本处，便可达到目的。

　　（3）移动及复制　　文本中的文字、句子、段落以及整页均能在该文件内移动或复制到任何地方。先将欲移动或复制的那整块文本做上记号，然后将光标放置在该块文本在文件中所出现的地方。此后，输入适当的命令以引导程序将该文本放置到新的地方。程序会自动调整周围的文本，以便给新的文本块让出空间。许多字本处理程序能将磁盘中文件的一部分或整个文件，复制到存储器中某个文件中。例如，标准的书信开头及称呼便能放置在文档中，可根据需要复制到新的文件中。

　　（4）搜索及替换　　搜索及替换功能可实现对选定的字母、单词或短语的某些部分或整体进行有选择性的修改。计算机在整个文件中逐字地进行搜索，以便查找出指定的修改文件。根据操作员所提供的不同指令，计算机或是停止在每一指定的更改处，让操作员对字符进行更改，或是用所需的更改内容自动替换原文本。

　　（5）列对齐　　在许多文字处理程序中，可运用自动列对齐程序对文本或数据进行列对齐。许多文字处理软件包的特定的功能，就是围绕小数点对数字自动进行列对齐，并能进行行和列的算术运算。例如，文字处理软件可对出现在文本中的数字表格进行行和列的数字归总，以便校对算术运算的结果。

（6）开头和结尾　像文件的每一页的开头都会出现的那种（具有固定格式）的页首那样的文本块，一旦一次性输入后，便可自动地复制到后续的各页中。这种功能一般用于识别文件的页码，以及自动进行页码累计计算。在进行某些文字处理时，其能跟踪下角标注并将其置于文件中适当的页码内。

（7）拼法检查　许多文字处理软件包得到电子词典及其处理程序的支持，因而其可自动地检查拼法上的错误。这些拼法检查器将整个文本中的字词，逐个地与词典中存储的字词进行比较。当发现有差异后，计算机记下那个不能识别的词，以便让操作人员进行修改。如果该词是对的，操作人员还可将其补充到词典的内容中。

（8）通信格式生成　经常保持着通信联系的人的名单和地址，可与文件分开存放。在打印时，这些人员名单和地址能与文件进行拼接组合，以形成私人性质形式的信件和备忘录。数据、通信地址、称呼及文本中其他各个不同的子块都能与文件拼接组合形成整体，这样对每个通信对象来说，其标准文件就不必每次都重打印了。

（9）打印格式　文件按照操作员选择的格式进行打印。通常的选项有如下几种：字体和行距、页码自动标记、右边界对齐，以及多份打印等。在对一个文件进行编辑时，可能会改变在原文本输入时所设定的页长。因此，文字处理程序常常具有重新调整每一页的长度的功能。自动地调整每一页的长度，将文本进行重新排版，以满足打印时所需的标准。

以上是专业的文字处理软件包的基本功能，而这些软件包又是商务或专业技术办公时的文本处理所要用到的。有些软件包的功能要比上述功能会更好一些，但也有一些软件包的功能会更差些。现有许多种文字处理软件可让用户根据自己的需要，选择适合自己文本生成、编辑、打印要求的软件包。

电子表格

电子表格单元格中的填写内容可在下列三类中选取其中的一种：
（1）名称　指明处于某一具体的行、列或者单元格里的数据属性。
（2）数据　表示需处理的财务或统计的数据。
（3）公式　指明对某一具体的单元格中的数据所需进行的运算。

例如，假设某公司计划利用表格分析六个月内的销售收入和生产成本。在建立表格的过程中，表格的第一行和第一列用来输入项目的名称这个信息。该信息所指的就是这一单元格的内容所处的行数和列数。然后，将实际的收入及支出之值输入到六个相邻的列中，每一列代表一个月的时间周期。

当原始数据输入到表格中，接着便会自动进行运算。例如，运算公式算出的是每月的收入和支出。但当该运算公式进入表格后，其通过计算机运算的结果，便是六个月内的总收入和总支出的比较情况。

作为表格的进一步应用，该实际的数值可用来规划未来的收支与支出。可先将收支相关的公式输入。然后，计算机便会自动地算出未来每个月的预期结果。这种预期规划结果会放置于在表格的另一列中。

表格生成软件包在财政计划、预算、预测等方面的运用提供了多种功能。许多电子表格都具备如下功能。

（1）自动运算　数据下载到表格中后，计算机便自动进行运算。无论是行还是列，都会扩展以容纳计算的结果。另一种方法是即将出现在某一单元格中的数值可以由一个公式来决定。该公式表示对处于某一单元格中的数值所要进行的数学运算。该运算公式也表示了运算的顺序及所指定的那个单元格的位置。而后，其计算结果的值便放在设计好了的那一单元格内。

行数与列数交结的那个数值便代表数据将要存放在的那一单元格。单元格的宽度可以变化，以适应所填入的数值的大小。例如，名称可以写入相邻的几个单元格内，而一个单元格又可容纳文本的一个整页。表格中的单元格还可缩减到小于标准值，以便在屏幕上可以看到表格的更大部分。

（2）运算模板　许多表格生成软件包都提供运算模板，即标准的运算公式。这些模板用于建立工作模板，以完成典型的数据处理任务。例如，标准模板可用来分析收支情况、规划现金流向、建立预算、预测走向、准备财务报告，以及其他常用的商务功能。这些模板（也就是财务模式）的优点是，当模板程序运行时，用户不必每次都要建立一个具体的表格。可将所选择的模板调出以生成工作模板，并在标准模板下允许数据输入。

（3）移动和复制　表格生成软件已设计成可以方便地进行表格的增加与减少。单一的单元格或是一片单元格区域均可从工作模板中所处的一个地方转移到另一地方。为了分析的方便和整个表格的视觉效果，软件在调整行和列的方面具有很大的灵活性。软件会根据原单元格所处的新位置，自动地调整公式。

（4）增加和删除　要在表格中增加新的行或列，或是删除一部分单元格是很容易的。这些功能可确保工作模板是保持动态的。当处于和移动及复制状态时，一旦行和列的计数改变，软件会将运算公式与所处的方格之间的关系自动地维持下来。

（5）滚动及多窗口显示　表格常常会变得比显示器屏幕还要大，特别是行数在25及其以上，列数在10及其以上的表格更是如此。在这些情况下，软件会让行和列滚动以便将工作模板上的某一指定部分展示在显示器屏幕上。当出现其他情况时，就应用多窗口显示技术。这种技术可让工作模板中两部分及其以上的较小部分同时出现在各自的窗口中，或者显示器中各自的部分屏幕上。

（6）报告打印　屏幕显示的内容可送至打印机作为硬拷贝输出。用户可打印工作模板上的某一部分，也可打印整张表格。大多数软件可将工作模板中的一部分复制到字处理程序中，以便表格可出现在报告中。

（7）图表　由于软件中的特殊处理程序可将数字转换成全标识的图表，因此出现在表格的单元格中的数据可以图形和图表的形式显示出来。图形和图表的大小可进行调整，以适合于在显示器上显示或在打印纸上打印。在某些显示器和打印机的输出上，可以看出各种不同的色彩，以示强调和醒目。

（8）文档下载　数据可以用以下两种方法放置于表格中的单元格内：一种是用户运用简单的编辑命令，将数据以插入、删除、修改或格式化等方法输入表格中。另一种是将数据从数据文档中下载到表格内。例如，目前保存在公司文档中的财务信息，可被自动地选取并输入表格中，而不必在每次分析时再将数据输入一遍。

电子表格可让用户进行大量的运算并能将其自动显示。大部分情况下，表格用于分析某一商务活动当前的财务状况，以及预测计划好的运行可能出现的结果。例如，用户可预备一份包括与商务情况有关的当前财务数据。然后，在某项计划、政策或发展趋势等对其的各种影响的假设条件下，有选择性地修改数据。这些假设以公式的形式出现，这些公式是表格中两个或以上的单元格内的数据之间的关系。而后，表格程序便进行必要的运算以预测未来的商务结果。商务经理在实施计划及政策前便能看到可能的结果。

Unit 5.4   Intelligent Technology

5.4.1   Text

Evolution of Intelligent Technology

During the last nine years, intelligent technology has evolved through three generations. The first was characterized by tools such as Prolog and OPS. Prolog is a simple backward-chaining environment (in which rules are linked in a top-down style), and OPS is a simple forward-chaining environment (in which rules are linked in a bottom-up style).

This led to the second generation of knowledge-engineering environment. Its goal was to model, prototype and construct knowledge systems.

The third generation was more concerned with operational requirement, performance, integration, database integration, connectivity and CASE-productivity tools.

All three generations were oriented toward generic problem solving and represented generalized implementation environments for the construction of expert system, knowledge bases and AI (Artificial Intelligence) application. This year people will witness the dawning of a fourth generation in which products are solution-oriented.

Some of the solution-oriented products are also seamlessly integrated with generic knowledge-engineering environments, permitting end users to add custom rules and objects, as well as predefined knowledge-based logic to address problems in manufacturing simulation.

Marketplace orientation is rapidly shifting toward solution oriented intelligent product environments. The fourth generation, which represents a new line of thinking and development time associated with problem solving.

There will always be a tremendous demand for generic knowledge-engineering environments from such areas as the aerospace, defense, manufacturing, finance and telecommunications industries. The number of potential problems far out weights the ability of the few companies in that field to build prepackaged intelligent solution. However, the key to intelligent technology becoming widespread and omnipresent lies in products that solve problems.

With respect to generalized knowledge-engineering environment, the industry has reached its third generation; however, in terms of intelligent application solution, it is at its first generation. In the knowledge-engineering environment, the thrust will be toward integration, performance, database connectivity, CASE productivity tools and adherence to standards.

It has become widely accepted that AI cannot be an island of automation. Traditional environments, such as MIS and DP departments, benefit from the integration of AI.

Another major theme will be the integration of knowledge bases with relational databases to deal with complex problems. Important in this area is the environment's ability to handle SQL (Structured Query Language), an industry standard for communicating with relational databases;

the X Window System, an industry standard for graphics and presentation management; and CLOS (Common Lisp Object System), an industry standard for object oriented programming and representation.

Maturing of AI

The year 1965 signaled the birth of a new branch of computer and cognitive science referred to as artificial intelligence (AI). AI development paralleled a three-stage process in hardware and software technology. The first stage of AI evolution was pure exploration. From the mid 1950s to the late 1970s, researchers from universities, industry, and government were interested in developing working models that tested ideas and theories about computational intelligence.

The second stage of the evolution of artificial intelligence was the prototype phase. By the early 1980s, AI technology matured. Tools for building AI programs were common. The first meeting of the American Association of Artificial Intelligence was held in the summer of 1980 at Stanford University. Japan launched its Fifth Generation project. Several European countries collaborated on the ESPRIT project. The UK initiated the Al program. The USA added its Strategic Computing Initiative.

The third and current stage of AI computer-system evolution is characterized by advances in AI integration, a transition from prototyping to incremental development and a shift back to AI on general-purpose computers.

By the mid 1980s, increasing numbers of AI systems were in use; users were looking for cost-effective solutions to their problems. AI tool builders were forced to integrate their AI software with databases, graphics, programming languages, and other applications. Concurrently, AI tools written in conventional programming languages, such as C, were becoming common.Although both of these trends spur AI integration, more important is the current trend to treat AI development as conventional software development.Integration has become part of the design process rather than an afterthought.

Occuring in parallel with the shift to conventional software development for AI applications was a move from prototyping to incremental development. Because AI tools provided better support for integration, it was possible to build and test increasingly complete versions of the final system. In the exploratory-prototype stages, the common practice was to build demonstration systems that were then later abandoned. Prototyping, however, continues to be important as a means of answering focused questions that facilitate the incremental development of software.

With the introduction of powerful workstations and advances in compilation technology, it became possible in the mid 1980s to implement efficient versions of symbolic programming languages on conventional computer systems. This development enabled many vendors and users to port their AI tools and applications onto general-purpose workstations.

Later, programming environments like those on Lisp machines were developed for general-purpose workstations. These conventional computing systems gave application builders easier access to conventional programming languages, networking, user interfaces, operating systems, and other application. Today mature AI technology is being delivered in conventional languages, such as C, along with traditional AI languages such as Lisp, Prolog, and Smalltalk. The AI research community continues to serve as a catalyst for new advances in both hardware and software and as a proving ground for the next generation of artificial intelligence products.

Fuzzy Logic

When Lofti Zadeh introduced the idea that computers could act on shades of gray information instead of the traditional clear cut yes no operations currently used, he was often mocked by the scientific community, sometimes for little more than the unscientific name he attached to his theory—"fuzzy logic"【1】. But Zadeh, a professor of computer science and electrical engineering at the University of California at Berkeley, is having the last laugh 25 years later【2】. The Japanese government and Japanese companies are introducing products based on his theory.

"The gain in acceptance has not been overnight, but there has been a quantum jump as a result of development in Japan," Zadeh says.He added that while Japan has only accounted for 5% of the academic research in fuzzy logic (much has been done in Europe, China and the Soviet Union), it is leading efforts to commercialize the technology, and other countries will have to follow suit.

Some of the products include a trading program for Yamaichi Securities Ltd.On the Tokyo market, which often responds to fads as well as the economic logic used in Wall Street, train controls, auto-focus cameras, elevator controls and automatic transmissions.

Pushing the field is the Japanese Ministry of International Trade and Industry, which formed the Laboratory for International Fuzzy Engineering (LIFE) in 1988. Its board of directors includes the presidents of Hitachi Ltd., Toshiba Crop., Fujitsu Ltd., Matsushits Electric Industrial Co.Ltd.and seven others, lending the project immediate respect with the expectation of commercial success.

Since the computer science community in the U.S.has been resistant to fuzzy logic, few indigenous corporations have any basis for developing products using it.Countries without an established computer science oligarchy, like Japan, were not burdened with such set concepts and explored fuzzy logic.

Japan, adopting it in a nearly faddish way (including adding the English word "fuzzy" to non-computer products such as toilet paper) set to making fuzzy logic a commercial success.

"Nissan patented a fuzzy logic transmission, Subaru has one, and Honda is experimenting with one", Zadeh said. GM and Ford and Chrysler will find themselves in a difficult situation (if they don't act on such a product). The same thing is going to happen in other fields. For example, Otis Elevator Co.recently sent a representative to a fuzzy logic conference because a Japanese elevator company already has a system for sale, Zadeh said.

Fuzzy logic can be in software, or software and hardware.Software sets include fuzzy predicates, fuzzy truth values, fuzzy probabilities and hedges, according to Zadeh. Fuzzy logic hardware carries current at differing levels of intensity to indicate values between 1 and 0.

One U.S.firm, Tongai Infralogic Inc. in Irvine, Calif., manufactures a fuzzy logic reduced instruction set computing chip for embedded use.The fuzzy sets for its use are in external memory. Programmers can use Infralogic's expert systems shell to compile data in C, assembly or microcode. Still, despite Infralogic's U.S.base, nearly all of its sales are to Japanese companies.

5.4.2 Specialized English Words

intelligent technology　智能技术
forward-chaining　向前链接
expert system　专家系统
AI（Artificial Intelligence）　人工智能
SQL（Structured Query Language）　结构化查询语言（一种与相关的数据库进行通信的工业标准）
general-purpose workstation　通用工作站
fuzzy predicates　模糊术语
fuzzy probabilities　模糊概率
backward-chaining　向后链接
CASE　计算机辅助软件工程
knowledge base　知识库
MIS and DP　管理信息系统与数据处理
X Window System　X Window 系统（一种图形和显示管理的工业标准）
CLOS（Common Lisp Object System）　一种面向对象的编程和表达的工业标准
user interfaces　用户接口
fuzzy truth value　模糊真值
assembly or microcode　汇编语言或微处理码

5.4.3 Notes

【1】When Lofti Zadeh introduced the idea that computers could act on shades of gray information instead of the traditional clear cut yes no operations currently used, he was often mocked by the scientific community, sometimes for little more than the unscientific name he attached to his theory—"fuzzy logic".

该句可谓一难长句。首先应该看出其主句句干是一主从复合句，主句为 he was often mocked by the scientific community, sometimes for little more than the unscientific name he attached to his theory—"fuzzy logic"，从句为 When Lofti Zadeh introduced the idea that computers could act on shades of gray information instead of the traditional clear cut yes no operations currently used。只不过在此主句中，还有一个由介词 for 引导的介词短语 sometimes for little more than the unscientific name he attached to his theory—"fuzzy logic"，其在句中做原因状语。然而，该介词短语又有一个修饰名词 the unscientific name 的定语从句 he attached to his theory—"fuzzy logic"，所以只一个主句就显得很长。此外，句中 When Lofti Zadeh introduced the idea that computers could act on shades of gray information instead of the traditional clear cut yes no operations currently used 是整句的时间状语从句，只不过它不是一个简单句，其中"that computers could act on shades of gray information instead of the traditional clear cut yes no operations currently used 就是 the idea 的同位语从句。根据如此分析，该句可翻译为："当 Lofti

Zadeh 提出计算机能作为灰度信息的一种斜度，而不是现在运用的传统的'是'与'非'运算的思想时，他常常遭到科学界的嘲笑，有时只不过是由于他给其理论起了'模糊逻辑'这样一个不科学的名字而已"。

【2】But Zadeh, a professor of computer science and electrical engineering at the University of California at Berkeley，is having the last laugh 25 years later.

在分析该句时应该看出句中的 a professor of computer science and electrical engineering at the University of California at Berkeley 为 Zadeh 的同位语，此处将其译为"身为加州大学伯克利分校的一名计算机科学和电子工程教授"。因此全句可译为："但是，身为加州大学伯克利分校的一名计算机科学和电子工程教授，Zadeh 在 25 年之后却笑到了最后"。

5.4.4 Translation

智 能 技 术

智能技术的发展

在过去的九年间，智能技术的发展已前后经历了三代。第一代是以 Prolog 和 OPS 这一类工具为特征的。Prolog 是一简单的向后链接环境（其规则为以自上而下的方式进行链接），OPS 是一简单的向前链接环境（其规则为以自下而上的方式进行链接）。

由这第一代智能技术便引出了其第二代，第二代智能技术为知识工程环境。其目标是得出知识系统的数学模型、物理原型和实际结构。

而第三代智能技术则更多的是关系到操作要求、性能、集成、数据库的集成、连接性和计算机辅助软件工程的生产率这样一些工具。

发展所有这三代智能技术是为了解决一些带有普遍性的问题，其代表专家系统的结构、知识库和人工智能应用抽象化的实施环境。今年，人们将会看到第四代智能技术的曙光，这一代智能技术的产品着眼于实际问题的解决。

某些面向解决实际问题的产品也与一般知识工程环境进行无缝地连接，可令终端用户在其内添加用户自己规则和要求。此外，还可预先定义基于知识的逻辑，以解决制造业仿真中的问题。

市场的方向也很快地转向了面向智能产品环境问题的解决方面。代表新的思维和新的发展时代的第四代产品，是与问题的解决相关联的。

宇航、国防、制造、金融和电信等部门，将一直对一般知识工程环境有相当大的需求。其潜在的需求数量，远远超出在此领域中拥有创建预包装的智能解决方法为数不多的那几家公司的能力。然而，智能技术能广为推广和深入各行各业的关键，就在于其能提供解决问题的产品。

就抽象化知识工程环境而言，已经达到了第三代工业化的水平。但就智能应用解决实际问题而言，尚处在第一代的水平。在知识工程环境中，其奋斗目标将是向集成、性能、数据库连接性，以及计算机辅助软件工程的生产率等这样一些工具，以及如何符合标准化的要求的方向前进。

人工智能不可能成为自动化领域中的孤岛，这一观点已广为人们所接受。另外一些传统的环境，如管理信息系统和数据处理部门，也从人工智能的集成中大为受益。

另一个重要研究课题是知识库与相关的数据库的集成，进而用以处理复杂的问题。在这个领域中，重要的是此环境的各种能力，即处理 SQL（结构查询化语言，一种与相关的数据库进行通信的工业标准）的能力，处理 X Window 系统（一种图形和显示管理的工业标准）的能力，以及处理 CLOS（一种面向对象的编程和表达的工业标准）的能力。

人工智能技术的成熟

1965 年标志着计算机和认知科学一个新分支的诞生，这就是通常所说的人工智能。在硬件与软件技术方面，人工智能经历了三个发展阶段，第一阶段是纯粹的探索阶段。从 20 世纪 50 年代中期到 70 年代末期，大学、工业界和政府研究人员的兴趣在于开发测试计算智能的想法和理论的工作模型。

人工智能发展的第二阶段是原型阶段。到 20 世纪 80 年代初，人工智能技术已趋成熟。编写人工智能程序的工具很普遍。1980 年的夏天，全美人工智能协会的第一次会议在斯坦福大学召开；日本开始了第五代的研究项目；数个欧洲国家合作进行 ESPRIT 计划；英国发起了"阿尔维"计划；美国也有了"战略计算开启"计划。

人工智能计算机系统发展的第三阶段，也就是当前的这个阶段，是以人工智能的集成化的进步为特征的，这种进步既是由原型产生向其增量开发的一种转换，又将人工智能返回到通用的计算机的运用上了。

到 20 世纪 80 年代中期，越来越多的人工智能系统被使用，当时用户在寻求解决其实际问题的经济有效的办法。人工智能工具的开发者们也被迫将其人工智能软件与数据库、图形绘制、编辑语言和其他应用集成起来。与此同时，用常规编程语言，如 C 语言编写人工智能工具也变得普遍起来了。尽管这两种倾向刺激了人工智能的集成化，但当前更重要的是将人工智能的开发作为常规的软件开发。集成化已成为设计过程的一部分，而不是事后再去进行补救。

与人工智能应用转到常规软件开发的同时，所出现的是从原型产生向其增量开发方向转移。由于人工智能工具为集成化提供了更好的支持，因此其能为最终的系统编制和测试逐步完整的版本。在原型的探索阶段，常见的做法是先绘制出展示系统，然后再将其废弃。然而，原型将继续是回答实现软件的增量开发这样一个焦点问题的重要方法。

随着功能强劲的工作站的出现和编辑技术的进步，20 世纪 80 年代中期，在常规计算机系统上实现符号编程的高效版本已成为可能。这一发展，使得许多供应商和用户能将其人工智能工具和应用程序移植到工作站上。

此后，可用于通用工作站的、类似于 Lisp 计算机上的那些编程环境被开发出来了。这些常规的计算机系统，可使应用开发人员更容易运用常规的编程语言、互联网、用户接口、操作系统和其他应用程序。今天，成熟的人工智能技术可以用常规语言（如 C 语言），以及传统的人工智能语言（如 Lisp，Prolog 和 Smalltalk 语言）一起提供。人工智能研究界继续起着硬件和软件的催化作用，并将成为下一代人工智能产品的基地。

模糊逻辑

当 Lofti Zadeh 提出计算机能作为灰度信息的一种斜度，而不是现在运用的传统的"是"与"非"运算的思想时，他常常遭到科学界的嘲笑，有时只不过是由于他给其理论起了个"模糊逻辑"这样一个不科学的名字而已。但是，身为加州大学伯克利分校的一名计算机科学和电子工程教授，Zadeh 在 25 年之后却笑到了最后。日本政府与日本公司正在根据他的理论推出相应的产品。

Zadeh 说："要人们接受它不是一夜之间就能实现的，但是日本方面对模糊逻辑的开发却产生了量的跃变。"他还说，日本在模糊逻辑的学术研究方面只占 5%（欧洲、中国和苏联已做了许多研究工作），但它在将该技术商品化方面却起着领头羊的作用，而其他国家也势必会随之仿效。

有些产品包括 Yamaichi 证券公司的一个交易程序，其在东京股票市场上是常用的工具。此外，还有些产品可运用于华尔街的经济逻辑规律的分析、列车的运行控制、自动聚焦照相机、电梯控制，以及自动传输等。

大力推进此项领域开发的是日本的通产省，其在 1988 年成立了国际模糊实验室（简称 LIFE）。该实验室的董事会成员包括日立、东芝、富士通、松下电气及其他七家公司的总裁，这有助于在商业上获得成功的希望得以实现。

由于美国的计算机科学界对模糊逻辑采取抵制态度，因而几乎没有纯粹的美国本土公司具有开发利用此项技术的产品的任何基础。没有建立起像日本那样的计算机科学的垄断体制的国家，是承担不了采用这种概念的重担和研究模糊逻辑的。

日本以几乎疯狂的方式采纳模糊逻辑（甚至对非计算机产品，如手纸，也加上 fuzzy 这个英文字），以使模糊逻辑获得商业上的成功。

Zadeh 说："日本尼桑汽车公司拥有一项模糊逻辑传输的专利，Subaru 公司也有一项，本田公司也有一项专利正在申请之中。"Zadeh 称，如果再不研制这类产品，美国的通用汽车、福特公司和克莱斯勒公司，会发现他们将处于困难的境地。同样的情况还将会在其他领域中也发生。例如，美国的 Otis 电梯公司最近派了一名代表参加模糊逻辑会议，因为一家日本电梯公司已经在销售（采用模糊逻辑的）系统了。

模糊逻辑可以放在软件中，也可以同时放在软件和硬件之中。据 Zadeh 说，软件组包括模糊述语、模糊真值、模糊概率和隔离方法。模糊逻辑硬件中流通着不同强度的电流，以在 1 或 0 之间表示其数值。

一家设在美国加州欧文市的 Tongai Infralogic 公司，已制造出一种模糊逻辑，其精简了指令组的计算芯片，因而可按嵌入式方式使用。为了便于运用，该模糊逻辑集存放在外存设备中。程序员可采用 Infralogic 公司的专家系统的外壳，用 C 语言、汇编语言或微处理码对数据进行编译。此外，尽管 Infralogic 公司的总部在美国，但几乎它的所有产品均销往日本公司。

Unit 5.5 Neural Nets

5.5.1 Text

Neural Nets Arrive

Neural networks employ a parallel processing approach rather than the procedural, or linear, design of conventional systems. The parallel design makes neural networks particularly adept at analyzing problems with many variables. By simultaneously considering numerous factors, neural networks map a mode, which encompasses acceptable solutions to a given problem.

The model for the processing method on which neural networks are based is the human brain. Neural networks mimic, in a very crude way, some functions of the brain.

Virtually all commercial neural network applications are restricted to the research laboratory—for now. The full potential of neural networks will not be tapped until parallel processor systems become more affordable 【1】.

Most commercial products available are based on the C language and require some familiarity with C programming as well as neural networking theory.

Generally, neural network components are arranged in a linear fashion comprising three basic, interconnected layers: an input layer, "hidden" layer, and an output layer. Groups of data, or inputs, are processed by the hidden layer, with the result being one or several outputs, or solutions, to a particular problem. The hidden layer is known as such because, unlike the input and output layers, the multiplying and summing operations that take place there are fixed and unavailable for adjustment.

The relative strengths, or mathematical values, of the various connections that transfer data from layer to layer are called weights Weights are crucial to neural network processing. It is through repeated adjustment of those weights that the network "learns"【2】. Application developers must "train" a neural network by feeding it a number of data samples before the system can produce solutions independently.

The methodologies by which neural networks learn, of which there are at least 14, are known as network paradigms. Much of the current neural network effort in research labs is in simply determining which paradigms are best matched to specific problems.

Hybrid applications are not only possible, they are desirable. Virtually all neural network applications can benefit from links with conventional algorithmic methods or expert systems, especially to "preprocess" data. Neural networks can also be combined with other technologies such as image processing. The pattern recognition and prediction capabilities of neural networks make them suitable for a number of manufacturing applications, including modeling complex, dynamic process control systems. Neural networks can also efficiently monitor manufacturing processes. Machine tool diagnostics is another area in which existing technologies can benefit from neural networks.

It will take time for a neural network to catch on outside the laboratory 【3】. The software is fairly cheap, but it is not easy to get the hang of neural networking.

It might take six months to come up with prototypes and another six months to develop workable production applications.

Glamour and Glitches

So far, the only large-volume application for neural networks is in high-speed modems. But maybe in less than two years, neural networks will be the tool of choice for analyzing the contents of large data bases.

Neural networks are being tried out in a variety of applications other than modems. For example:

A system reads handwriting and converts it into ASCII characters as though the information had been entered through a keyboard instead of a digitizing pad.

A product can recognize and classify complex visual images, a feature useful in applications such as industrial parts inspection and autonomous robotics.

A low cost software product lets developers experiment with neural network architecture.

Neural networking is an information processing approach entirely different from the conventional algorithmic programming model. Neural networks (neuro-computers) are trained for a specific task in much the same way a human would be. That is, given a set of examples of the kinds of inputs it will receive and the sort of output required, a neurocomputer will develop its own problem solving algorithm.

Since neuro-computers function essentially by recognizing patterns, they are well suited for tasks that conventional computers cannot do, for example, looking at masses of marketing and sales data to discern buying patterns of a product. This is what expert systems do. A neural network-based expert system does not need rules, programming, or high-priced knowledge-engineering services.

Nestor Inc. in Providence RI, is the patriarch of the neural industry. There are now at least 125 companies in major neural networking development research. But today's buyers are investing in research and development, not in a mature technology. Several years will have to pass before neuro-computers join main stream computing.

Neuro-computer architecture is modeled on current understanding of the way the human brain encodes and processed information. Nodes in a neural network, called processing elements, model individual nerve cells, or neurons, in the human nervous system. In the brain, each neuron receives electrochemical impulses from thousands of neurons. If the sum of the signals received is strong enough, the neuron sends it on to another nerve cell or to a muscle or organ. Laboratory scientists are now trying to determine how the organic network of nerve cells is arranged and what happens within a neuron when learning takes place.

Neural network simulation software lets developers work with some of the dozens or so best-known theories of brain organization and learning. They can experiment with processing elements and the strengths of signals between the elements. The brain is a massively parallel processor, working on many input at once; neuro-computers are also inherently parallel. But most currently available products are sofeware packages of add-in boards that simulate neural networks in conventional serial computers.

The processors in a neural network, whether implemented in hardware or software, are relatively simple. Neuro-computers work in parallel, with several processing elements at once handling a single data item. Learning is distributed among processing elements which leads to unusual robustness in neural networks. Neuro-computers are inherently fault tolerant; individual processing elements can malfunction without compromising the entire system, just as individual cells in the brain can die with no change in a person behavior.

Typically, processing elements are arranged in layers. The lowest layers look at general features of a pattern, and succeeding layers are increasingly specific until at last there is a layer whose output is sufficiently refined to be useful.

Neuro-computing finds its roots in technologies such as signal processing and pattern recognition, and dates back to the 1940s and a paper written by neurophysiologist Warren McCulloch and mathematician Walter Pitts. In 1951, Harvard University undergraduate Marvin Minsky built a learning machine out of vacuum tubes and motors based on the McCulloch-Pitts neuron model, a model that to this day, forms the basis of most work in neural networking 【4】. (Minsky is one of the founding fathers of artificial intelligence.)

Given the current lack of knowledge about which network configurations best suit which kinds of tasks simulating neural networks in software in the logical choice, but eventually, we will see neural networks on a chip.

The implications of neuro-computing in fields such as manufacturing and process control are many and varied. Neural networks could also handle data base management tasks such as eliminating duplications from a mailing list.

One of the most interesting product categories is an adaptive expert in a financial services firm to make recommendations concerning mortgage applications. The system is different from a conventional expert system in which rules are developed from a set of examples, in that there are no predetermined relationships between the variables 【5】.

However, the potential of neural expert systems is impressive.Neurocomputer-based systems that recognize people's voices or faces or palm prints are becoming possible and neural-equipped watch-man robots should be able to differentiate between animals and human intruders in a factory, to see fire or flood, and to decide on appropriate actions to take.

Continuous speech-recognition systems will make possible voice typewriters that take vocal dictation and turn out printed memos. But first, special purpose hardware will have to be developed. In addition, the equations that determine the strength of connections between processing elements are both difficult to select and very critical.

In the future, users may be able to buy neural network shells that incorporate answers to many of these questions. But at the moment, each system is proprietary and custom-written. User companies with neurocomputer applications underdevelopment and the network vendors working with them are usually sworn to secrecy.

5.5.2 Specialized English Words

neural network   神经网络
C language   C 语言
weight   （加）权（重）
preprocess data   预处理数据
pattern recognition   模式识别
modeling complex   建模复杂的
machine tool   机床
modem   调制解调器
signal-processing   信号处理
parallel processing approach   并行处理方式
C programming   C 语言编程
expert system   专家系统
image processing   图像处理
prediction capability   预测能力
dynamic process control system   动态过程控制系统
diagnostic   故障诊断
robustness   鲁棒性
adaptive expert   自适应专家系统

5.5.3 Notes

【1】The full potential of neural networks will not be tapped until parallel processor systems become more affordable.

需要注意的是，该句的时间状语从句是由从属连词 until 引出的，而主句中的谓语动词又是否定形式，因此翻译时不能照字面译为"一直到并行处理器系统能为更多的人买得起时，神经网络的全部潜能都不会得到人们的认可"，这样的译文会使读者不知所云，至少是没有将其意翻译得确切，使人明白易懂。为此，应将其主句中的谓语动词翻译成肯定形式，即"只有当并行处理器系统能为更多人买得起时，神经网络的全部潜能才会得到人们的认可。"

【2】It is through repeated adjustment of those weights that the network "learns".

应该看出该句为一个强调句型，因此应将其强调的语气翻译出来，即该句应译为："正是通过对那些权重的反复调整，神经网络才得以具备'学习'功能。"

【3】It will take time for neural networks to catch on outside the laboratory.

在翻译本句时，应该注意其中的动宾结构 take time 的用法，其意为"需要花费一些时间"。什么东西需要花费一些时间呢？就是 to catch on outside the laboratory（神经网络要想走出实验室），但由于该动词不定式短语做主语时太长，会显得头重脚轻，因此用先行词"it"替代它先行做主语。看懂了这一点，该句就不难翻译了。其应译为："神经网络要想走出实验室还有待时日。"

【4】In 1951，Harvard University undergraduate Marvin Minsky built a learning machine out of vacuum tubes and motors based on the McCulloch-Pitts neuron model，a model that to this day，forms the basis of most work in neural networking.

在翻译该句时重要的是要看出句中的 a model，并不是其前面的 pitts neuron model，若将其看作是同一个的话，那么 a model that to this day，forms the basis of most work in neural networking 这个 "直至今天便形成了神经网络中大多数工作的基础的模型"，便成了 the McCulloch-pitts neuron model，即 "McCulloch-Pitts 的神经模型" 了，这显然是不对的。这个模型应该就是 a learning machine，即 "一台学习机"。这是从逻辑上看是如此，若从英语语法上看也是如此。如果这两个 model 是同一个的话，那么其第二次出现时就应该是 the model 的形式，即其不能用不定冠词 a 而应用定冠词 the 来修饰。将以上情况弄清楚后，该句便不难翻译了。其应译为："1951 年，哈佛大学的研究生 Marvin Minsky，根据 McCulloch-Pitts 的神经模型，采用真空管和电动机制作出了一台学习机，这便是最初的学习机的模型，直至今天便形成了神经网络中大多数工作的基础。"

【5】The system is different from a conventional expert system in which rules are developed from a set of examples，in that there are no predetermined relationships between the variables.

在翻译该句时应该看出句中的 in which rules are developed from a set of examples 是修饰 a conventional expert system 的定语从句，而 in that there are no predetermined relationships between the variables 则是修饰 The system 的定语从句，只不过其二者离得较远，不易看出，要学会判别这种限定性的定语从句与其被修饰的名词相距较远的情形。清楚了这一点，该句就不难翻译了。其可译为："这种系统与往常的专家系统是不同的，后者的规则是用一组范例提炼出来的，而这种自适应专家系统中的变量之间无预定的关系"。

5.5.4 Translation

神 经 网 络

神经网络出现了

神经网络采用并行处理方式，而不是传统系统的程序化，即线性化的设计。这种并行的设计使得神经网络特别擅长于分析具有许多变量的问题。通过同时考虑多方因素，神经网络勾画出一种模式，其可囊括对某一问题的所有可接受的解决方法。

神经网络处理问题的模式是基于人脑的思维方式。神经网络粗略地模拟人脑的某些功能。

迄今为止，实际上所有具有商业价值的神经网络应用均仍处于研究性的实验之中。只有当并行处理器系统能为更多人买得起时，神经网络的全部潜能才会得到人们的认可。

现在能买到的产品都是基于 C 语言的，其需要熟练的 C 语言编程和神经网络原理的知识。

一般说来，神经网络的各个单元按线性格局排列，由以下三个相互连接的基本层面组成：输入层、"隐蔽" 层和输出层。输入层也就是一组数据，其经隐蔽层处理后便产生一组或几组输出，这也就是对具体问题的解。隐蔽层之所以有这样的称谓，是因为其与输入层和输出层不同，在此处进行的乘法和加法运算是固定的，是不能进行调整的。

各层面之间传递数据的各种互连的相对强度，也就是数学值，称为权。权对神经网络的处理是至关重要的。正是通过对那些权重的反复调整，神经网络才得以具备 "学习" 功能。

在系统能独立地解决问题之前，应用程序的开发人员必须给神经网络输入一些数据样本，以对其进行"训练"。

神经网络学习所采用的方法模式，至少有14种之多，其便称之为网络范例。目前在从事研究的实验室里，神经网络的许多工作只是简单地确定哪些范例与某些具体问题最为匹配。

（常规系统与神经网络的）混合应用不仅是可能的，也是人们所希望的。实际上，所有的神经网络应用，都能从其与运用常规算法的方法或专家系统的结合获得好处，尤其是对"预处理"数据来说更是如此。神经网络也能与其他的技术，如图像处理等相结合。

神经网络的模式识别和预测能力，使得其很适合与许多生产制造方面的应用，其包括各种数学建模复杂的动态过程控制系统。神经网络还能有效地监控生产制造过程。机床的故障诊断是又一个现有技术可以从神经网络获取好处的领域。

神经网络要想走出实验室还有待时日。软件倒是相当便宜，但要得到实用却并非易事。也许还需六个月的时间才会有神经网络原型问世，而要开发出可用于生产实际的运用软件，或许还得需要六个月的时间。

魅力与难点

迄今为止，神经网络唯一大量的应用是在高速的调制解调器中。但是，可能要不了两年的时间，神经网络将是分析大型数据库中的选中内容的得力工具。

神经网络正在各种各样的实际应用中进行试验研究，而并非只局限于在调制解调器之中的运用，例如：

有一种系统能读出手写字体，并将其转换成 ASCII 字符，就好像信息是通过键盘而不是经数字化的路径输入的。

又有一种产品能识别复杂的视觉图像，并对其进行分类，这种特性在机器零件的检验和自动化机器人这一类的应用中是很有用的。

还一种低成本的软件产品，可令开发人员用神经网络结构进行试验。

神经网络是一种与常规的算法编程模式截然不同的信息处理方法。神经网络（神经计算机）可接受针对具体任务进行训练，这很像是人在接受训练的那种方式一样。也就是说，人们给它一组其将接收的各种输入的样本，以及需要它输出的种类，神经计算机将自行得出解决问题的算法。

由于神经网络计算机基本上是通过模式识别进行工作的，因而其很适合做那些常规计算机不适合做的工作，例如，观察大量的销售数据，从中找出一种产品的购买模式，而这正是专家系统所要做的事情。基于神经网络的专家系统不需要规则、编程或高价的知识工程服务。

位于罗德岛州 Providence 市的 Nestor 公司是神经网络工业的鼻祖。现在从事重要神经网络开发研究的公司至少有 125 家。但今天的买主是在对研究与开发进行投资，而不是对成熟的技术进行投资。还得需要几年的时间，神经网络计算机才能成为计算机市场的主流。

神经网络计算机的结构，是按照目前对人脑编码和处理信息方法的认识为设计模型的。神经网络中的节点，也就是处理单元，是以人的神经系统中的单个神经细胞（即神经元）为模型的。在人的大脑中，神经细胞从成千上万的神经中接收到电化学脉冲。如果接收到的信号之和已足够之强，则神经元便将其发送给另一个神经细胞或者肌肉或者器官。实验室里的科学家们目前正在力图弄清神经细胞的有机网络是如何安排的，以及在进行学习时神经中将会发生什么情况。

神经网络模拟软件让开发人员用几十种人们熟知的人脑组织和学习的理论进行研究工作。他们能根据处理单元和这些单元之间信号的强度进行试验。人脑是一个大容量的平行处理器，一次将接收到大量的输入信号，而神经网络计算机本质上也是并行运行的。但是，大多数目前能够买得到的产品是扩充式的软件包，其采用常规的串行计算机模拟神经网络。

不管是用软件还是硬件，其实现的神经网络处理器都相对比较简单。神经网络计算机并行地工作，几个处理器同时处理某一个数据。学习是分散在多个处理单元上进行的，这便使得神经网络具有异乎寻常的鲁棒性。神经网络计算机在根本上就具有很好的容错性，如果个别的处理单元的功能出现故障，并不会连累整个系统，这就像人的大脑中的个别细胞可能会死亡，而人的行为举止并不会因此而有所改变。

通常处理单元是分为几层来安排的。最底下的那几层观察模式的一般特征，往上的各层其任务地逐步具体化，到最上面的那一层，其输出已足够精细而得出有用的结果。

神经网络计算起源于信号处理和模式识别这样一些技术，其可追溯到20世纪40年代的一篇由神经生理学家Warren McCulloch和数学家Walter Pitts合写的论文。1951年，哈佛大学的研究生Marvin Minsky，根据McCulloch-Pitts的神经模型，采用真空管和电动机制作出了一台学习机，这便是最初的学习机的模型，直至今天便形成了神经网络中大多数工作的基础。(Minsky也便是人工智能的创始人之一。)

尽管目前还不知道哪一种网络结构最适合哪一种任务，但是最终我们还是会看到做在一个芯片上的神经网络。

神经计算技术在诸如加工制造和过程控制等领域中的用途是很广泛的，并且其形式也是多种多样的。神经网络也能处理数据库管理任务，如在邮寄包裹单上消除重复。

一个最为有趣的产品类别是金融服务公司的自适应专家系统，其可作出有关抵押申请的推荐意见。这种系统与往常的专家系统是不同的，后者的规则是用一组范例提炼出来的，而这种自适应专家系统中的变量之间无预定的关系。

然而，神经专家系统的潜力是惊人的。能识别人的声音或者面孔或者手掌纹路的基于神经计算机的系统正在变成现实。配有神经网络的警卫机器人将能区别进入工厂的是动物还是闯入的人，能监视火警或水警，还能决定采取相应的行动。

连续语音识别系统将实现声音打字机，即输入声音听写和输出打印好的备忘录，但其前提是应先开发出专用的硬件。此外，确定处理单元之间连接强度的关系式，既难以确定又非常关键。

不久的将来，用户也许就可以购买到神经网络的外壳，其装有这些问题的答案。但就目前而言，每个系统都是专有的和定制编写的。尚未充分开发神经网络计算机应用的用户公司正与其供应商在一起商谈保密事宜。

Reference

[1] 戴文进主编. 专业英语阅读与翻译. 北京：航空工业出版社，1998.
[2] 戴文进主编. 自动化专业英语. 武汉：武汉理工大学出版社，2001.
[3] 戴文进著. 科技英语翻译理论与技巧. 上海：上海外语教育出版社，2002.
[4] 戴文进主编. 电气工程及其自动化专业英语. 北京：电子工业出版社，2004.
[5] 戴文进译著. 电机原理与设计的 MATLAB 分析. 北京：电子工业出版社，2006.
[6] Jimmie J.Cathey.Electric Machines：Analysis and Design Applying MATLAB.McGraw-Hill，2001.

反侵权盗版声明

电子工业出版社依法对本作品享有专有出版权。任何未经权利人书面许可，复制、销售或通过信息网络传播本作品的行为，歪曲、篡改、剽窃本作品的行为，均违反《中华人民共和国著作权法》，其行为人应承担相应的民事责任和行政责任，构成犯罪的，将被依法追究刑事责任。

为了维护市场秩序，保护权利人的合法权益，我社将依法查处和打击侵权盗版的单位和个人。欢迎社会各界人士积极举报侵权盗版行为，本社将奖励举报有功人员，并保证举报人的信息不被泄露。

举报电话：（010）88254396；（010）88258888

传　　真：（010）88254397

E-mail：dbqq@phei.com.cn

通信地址：北京市海淀区万寿路173信箱
　　　　　电子工业出版社总编办公室

邮　　编：100036